Ecology: Concepts and Applications

Ecology: Concepts and Applications

Jaxon Pine

Larsen & Keller
www.larsen-keller.com

Ecology: Concepts and Applications
Jaxon Pine
ISBN: 978-1-64172-412-8 (Hardback)

⊟ Larsen & Keller

Published by Larsen and Keller Education,
5 Penn Plaza,
19th Floor,
New York, NY 10001, USA

Cataloging-in-Publication Data

Ecology : concepts and applications / Jaxon Pine.
    p. cm.
Includes bibliographical references and index.
ISBN 978-1-64172-412-8
1. Ecology. 2. Biology. 3. Environmental sciences. I. Pine, Jaxon.
QH541 .E26 2020
577--dc23

For more information regarding Larsen and Keller Education and its products, please visit the publisher's website www.larsen-keller.com

TABLE OF CONTENTS

Preface		**VII**

Chapter 1 Introduction — **1**

- Ecology — 1
- Terrestrial Ecology — 7
- Microbial Ecology — 8
- Forest Ecology — 14
- Population Ecology — 15
- Behavioral Ecology — 30
- Applied Ecology — 31
- Ecological Succession — 33

Chapter 2 Ecosystem and its Types — **36**

- Components of Ecosystem — 38
- Forest — 41
- Desert Ecosystem — 60
- Aquatic Ecosystem — 65
- Grasslands — 67
- Taiga — 72
- Tundra — 88
- Montane Ecosystems — 93

Chapter 3 Ecology of Plant — **95**

- Plant Ecology — 95
- Plant Community — 96
- Plant Perception — 97
- Plant Stress — 103
- Plant Stress Measurement — 104

Chapter 4	Human Ecology	111
	• Cold and Heat Adaptations in Humans	112
	• Coupled Human–environment System	115
	• Urban Ecology	117
	• New Urbanism	126
	• Anthropocene	132

Chapter 5	Ecology of Animals	136
	• Animal Ecology	136
	• Animal Diversity	147

Chapter 6	Aquatic Ecology	151
	• Freshwater Ecosystem	154
	• Lake Ecosystem	156
	• River Ecosystem	167
	• Marine Ecosystem	183

Chapter 7	Systems Ecology	189
	• Ecological Systems Theory	189
	• The Ecological Laws of Thermodynamics	192
	• Energy Flow in Ecosystem	195
	• Nutrient Cycle	198
	• Food Chain	204
	• Food Web	205
	• Ecological Pyramid	214
	• Ecological Buffers	217

Permissions

Index

PREFACE

This book is a culmination of my many years of practice in this field. I attribute the success of this book to my support group. I would like to thank my parents who have showered me with unconditional love and support and my peers and professors for their constant guidance.

Ecology is a biological branch that is concerned with the study of the interactions between organisms and their environment. It primarily focuses on the biotic and abiotic components of the organisms' environment. Ecology studies the distribution, biodiversity, biomass and populations of organisms. It also focuses on the competition and cooperation among species. Concepts from this field are often applied in areas such as wetland management, conservation biology, city planning, community health, applied science, human social interaction, etc. It is also applied in natural resource management such as forestry, agroforestry and agriculture. Ecology works to explain the movement of energy and materials via living communities, life processes, adaptations and interactions, the successional development of ecosystems, and the distribution of organism's biodiversity in the environment. This book is a compilation of chapters that discuss the most vital concepts in the field of ecology. Such selected concepts that redefine this field have been presented herein. The textbook is appropriate for students seeking detailed information in this area as well as for experts.

The details of chapters are provided below for a progressive learning:

Chapter – Introduction

The branch of biology that deals with the study of interactions among the organisms and their biophysical environment is known as ecology. Some of the branches of ecology are terrestrial ecology, microbial ecology, forest ecology, population ecology, behavioral ecology and applied ecology. This is an introductory chapter which will briefly introduce all these branches of ecology.

Chapter – Ecosystem and its Types

The community of living organisms along with the non-living components of their environment is called an ecosystem. The major types of ecosystem are desert ecosystem, aquatic ecosystem, grasslands and montane ecosystems. This chapter has been carefully written to provide an easy understanding of these types of ecosystem.

Chapter – Ecology of Plant

Plant ecology is concerned with the study of relationships between plants and their physical and biotic environment. Some of its focus areas include plant community, plant perception and plant stress measurement. The topics elaborated in this chapter will help in gaining a better perspective about plant ecology.

Chapter – Human Ecology

Human ecology refers to the study of relationship between human and their social, natural and built environments. Urban ecology, new urbanism, anthropocene, etc. are some of the aspects that fall under its domain. This chapter closely examines these key concepts of human ecology to provide an extensive understanding of the subject.

Chapter – Ecology of Animals

Animal ecology is concerned with the relationships of animals with their environments. It also studies the consequences of these relationships for evolution, population growth and regulation. The topics elaborated in this chapter will help in gaining a better perspective about these areas of animal ecology.

Chapter – Aquatic Ecology

Aquatic ecology refers to the study of plants and animals living in water and their surroundings. Aquatic ecosystem, freshwater ecosystem, lake ecosystem, river ecosystem and marine ecosystem are a few branches of aquatic ecology. This chapter has been carefully written to provide an easy understanding of these branches of aquatic ecology.

Chapter – Systems Ecology

Systems ecology is referred to the field of ecology that deals with the study of ecological systems. It focuses on the study of the ecological pyramid, food web, food chain, nutrient cycle, ecological buffers, energy flow, ecological laws, etc. All these diverse aspects related to systems ecology have been carefully analyzed in this chapter.

Jaxon Pine

Introduction

1

- **Ecology**
- **Terrestrial Ecology**
- **Microbial Ecology**
- **Forest Ecology**
- **Population Ecology**
- **Behavioral Ecology**
- **Applied Ecology**
- **Ecological Succession**

The branch of biology that deals with the study of interactions among the organisms and their biophysical environment is known as ecology. Some of the branches of ecology are terrestrial ecology, microbial ecology, forest ecology, population ecology, behavioral ecology and applied ecology. This is an introductory chapter which will briefly introduce all these branches of ecology.

Ecology

Ecology, also called bioecology, bionomics, or environmental biology is the study of the relationships between organisms and their environment. Some of the most pressing problems in human affairs—expanding populations, food scarcities, environmental pollution including global warming, extinctions of plant and animal species, and all the attendant sociological and political problems—are to a great degree ecological.

The word *ecology* was coined by the German zoologist Ernst Haeckel, who applied the term *oekologie* to the "relation of the animal both to its organic as well as its inorganic environment. Thus, ecology deals with the organism and its environment. The concept of environment includes both other organisms and physical surroundings. It involves relationships between individuals within a population and between individuals of different populations. These interactions between individuals, between populations, and between organisms and their environment form ecological systems, or ecosystems. Ecology has been defined variously as "the study of the interrelationships of organisms with their environment and each other," as "the economy of nature," and as "the biology of ecosystems."

Ecology had no firm beginnings. It evolved from the natural history of the ancient Greeks, particularly Theophrastus, a friend and associate of Aristotle. Theophrastus first described the interrelationships between organisms and between organisms and their nonliving environment. Later foundations for modern ecology were laid in the early work of plant and animal physiologists.

In the early and mid-1900s two groups of botanists, one in Europe and the other in the United States, studied plant communities from two different points of view. The European botanists concerned themselves with the study of the composition, structure, and distribution of plant communities. The American botanists studied the development of plant communities, or succession. Both plant and animal ecology developed separately until American biologists emphasized the interrelation of both plant and animal communities as a biotic whole.

During the same period, interest in population dynamics developed. The study of population dynamics received special impetus in the early 19th century, after the English economist Thomas Malthus called attention to the conflict between expanding populations and the capability of Earth to supply food. In the 1920s the American zoologist Raymond Pearl, the American chemist and statistician Alfred J. Lotka, and the Italian mathematician Vito Volterra developed mathematical foundations for the study of populations, and these studies led to experiments on the interaction of predators and prey, competitive relationships between species, and the regulation of populations. Investigations of the influence of behaviour on populations were stimulated by the recognition in 1920 of territoriality in nesting birds. Concepts of instinctive and aggressive behaviour were developed by the Austrian zoologist Konrad Lorenz and the Dutch-born British zoologist Nikolaas Tinbergen, and the role of social behaviour in the regulation of populations was explored by the British zoologist Vero Wynne-Edwards.

While some ecologists were studying the dynamics of communities and populations, others were concerned with energy budgets. In 1920 August Thienemann, a German freshwater biologist, introduced the concept of trophic, or feeding, levels, by which the energy of food is transferred through a series of organisms, from green plants (the producers) up to several levels of animals (the consumers). An English animal ecologist, Charles Elton, further developed this approach with the concept of ecological niches and pyramids of numbers. In the 1930s, American freshwater biologists Edward Birge and Chancey Juday, in measuring the energy budgets of lakes, developed the idea of primary productivity, the rate at which food energy is generated, or fixed, by photosynthesis. In 1942 Raymond L. Lindeman of the United States developed the trophic-dynamic concept of ecology, which details the flow of energy through the ecosystem. Quantified field studies of energy flow through ecosystems were further developed by the brothers Eugene Odum and Howard Odum of the United States; similar early work on the cycling of nutrients was done by J.D. Ovington of England and Australia.

Konrad Lorenz.

The study of both energy flow and nutrient cycling was stimulated by the development of new materials and techniques—radioisotope tracers, microcalorimetry, computer science, and applied mathematics—that enabled ecologists to label, track, and measure the movement of particular nutrients and energy through ecosystems. These modern methods encouraged a new stage in the development of ecology—systems ecology, which is concerned with the structure and function of ecosystems.

Areas of Study

Ecology is necessarily the union of many areas of study because its definition is so all-encompassing. There are many kinds of relationships between organisms and their environment. By organisms one might mean single individuals, groups of individuals, all the members of one species, the sum of many species, or the total mass of species (biomass) in an ecosystem. And the term environment includes not only physical and chemical features but also the biological environment, which involves yet more organisms.

In practice, ecology is composed of broadly overlapping approaches and further divided by the groups of species to be studied. There are many, for example, who specialize in the field of "bird behavioral ecology. The main approaches fall into the following classes.

Evolutionary ecology examines the environmental factors that drive species adaptation. Studies of the evolution of species might seek to answer the question of how populations have changed genetically over several generations but might not necessarily attempt to learn what the underlying mechanisms might be. Evolutionary ecology seeks those mechanisms. Thus, in the well-known example of the peppered moth, the populations in the industrialized English Midlands changed over generations from having wings coloured largely grayish white, peppered with black spots, to wings that were mostly blackish. The ecological mechanism involved predation—birds readily detected the light-coloured moths against the background of the tree trunks that industrial pollution had darkened, whereas the dark-coloured moths remained generally undetected.

A light gray peppered moth (Biston betularia) and a darkly pigmented variant rest near each other on the trunk of a soot-covered oak tree. Against this background, the light gray moth is more easily noticed than the darker variant.

Evolutionary ecology also examines broader issues, such as the observations that plants in arid environments often have no leaves or else very small ones or that some species of birds have helpers at the nest—individuals that raise young other than their own. A critical question for the subject is whether a set of adaptations arose once and has simply been retained by all species descended from a common ancestor having those adaptations or whether the adaptations evolved repeatedly because of the same environmental factors. In the case of plants that live in arid environments, cacti from the New World and euphorbia from the Old World can look strikingly similar even though they are in unrelated plant families.

Physiological ecology asks how organisms survive in their environments. There is often an emphasis on extreme conditions, such as very cold or very hot environments or aquatic environments with unusually high salt concentrations. Examples of the questions it may explore are: How do some animals flourish in the driest deserts, where temperatures are often high and freestanding water is never available? How do bacteria survive in hot springs, such as those in Yellowstone National Park in the western United States, that would cook most species? How do nematodes live in the soils of dry valleys in Antarctica? Physiological ecology looks at the special mechanisms that the individuals of a species use to function and at the limits on species imposed by the environment.

Behavioral ecology examines the ecological factors that drive behavioral adaptations. The subject considers how individuals find their food and avoid their enemies. For example, why do some birds migrate while others are resident? Why do some animals, such as lions, live in groups while others, such as tigers, are largely solitary?

Population ecology, or autecology, examines single species. One immediate question that the subject addresses is why some species are rare while others are abundant. Interactions with other species may supply some of the answers. For example, enemies of a species can restrict its numbers, and those enemies include predators, disease organisms, and competitors—i.e., other species. Consequently, population ecology shares an indefinite boundary with community ecology, a subject that examines the interactions between several to many species. Species abundances vary both from year to year and across the species' geographic range. Population ecology asks what causes abundances to fluctuate. Why, for example, do numbers of some species, typically birds and mammals, change perhaps threefold or fourfold over a decade or so, while numbers of other species, typically insects, vary tenfold to a hundredfold from one year to the next? Another key question is what limits abundance, for, without limits, species numbers would grow exponentially.

Biogeography is the study of the geographical distribution of organisms, and it asks questions that parallel those of population ecology. Some species have tiny geographical ranges, being restricted to perhaps only a few square kilometres, while other species have ranges that cover a continent. Some species have more-or-less fixed geographical ranges, while others fluctuate, and still others are on the increase. If a species that is spreading is an agricultural pest, a disease organism, or a species that carries a disease, understanding the reasons for the increasing range may be a matter of considerable economic importance. Biogeography also considers the ranges of many species, asking why, for example, species with small geographic ranges are often found in special places that house many such species rather than scattered randomly about the planet.

Community ecology, or synecology, considers the ecology of communities, the set of species found in a particular place. Because the complete set of species for a particular place is usually not known, community ecology often focuses on subsets of organisms, asking questions, for example, about plant communities or insect communities. A fundamental question deals with the size of the "set of species"—that is, what ecological factors determine how many species are present in an area. There are many large-scale patterns; for example, more species are present in larger areas than smaller ones, more on continents than on islands (especially remote ones), and more in the tropics than in the Arctic. There are many hypotheses for each pattern. Ecological factors also cause the diversity of species to vary over smaller scales. For example, though predators may be harmful to individual species, the presence of a predator may actually increase the number of species present in a community by limiting the numbers of a particularly successful competitor that otherwise might monopolize all the available space or resources.

The questions above are generally applied to species at the same trophic level—say, the plants in a community, or the insects that feed on the plants there, or the birds that feed on the insects there. Yet a different set of questions in community ecology involves how many trophic levels there are in a particular place and what factors limit that number.

Conservation biology seeks to understand what factors predispose species to extinction and what humans can do about preventing extinction. Species in danger of extinction are often those with the smallest geographic ranges or the smallest population sizes, but other ecological factors are also involved.

Ecosystem ecology examines large-scale ecological issues, ones that often are framed in terms not of species but rather of measures such as biomass, energy flow, and nutrient cycling. Questions include how much carbon is absorbed from the atmosphere by terrestrial plants and marine phytoplankton during photosynthesis and how much of that is consumed by herbivores, the herbivores' predators, and so on up the food chain. Carbon is the basis of life, so these questions may be framed in terms of energy. How much food one has to eat each day, for instance, can be measured in terms of its dry weight or its calorie content. The same applies to measures of production for all the plants in an ecosystem or for different trophic levels of an ecosystem. A basic question in ecosystem ecology is how much production there is and what the factors are that affect it. Not surprisingly, warm, wet places such as rainforests produce more than extremely cold or dry places, but other factors are important. Nutrients are essential and may be in limited supply. The availability of phosphorus and nitrogen often determines productivity—it is the reason these substances are added to lawns and crops—and their availability is particularly important in aquatic systems. On the other hand, nutrients can represent too much of a good thing. Human activity has modified global ecosystems in ways that are increasing atmospheric carbon dioxide, a carbon source but also a greenhouse gas, and causing excessive runoff of fertilizers into rivers and then into the ocean, where it kills the species that live there.

Methods in Ecology

Because ecologists work with living systems possessing numerous variables, the scientific techniques used by physicists, chemists, mathematicians, and engineers require modification for use in ecology. Moreover, the techniques are not as easily applied in ecology, nor are the results as precise as those obtained in other sciences. It is relatively simple, for example, for a physicist to measure gain and loss of heat from metals or other inanimate objects, which possess certain constants of conductivity, expansion, surface features, and the like. To determine the heat exchange between an animal and its environment, however, a physiological ecologist is confronted with an array of almost unquantifiable variables and with the formidable task of gathering the numerous data and analyzing them. Ecological measurements may never be as precise or subject to the same ease of analysis as measurements in physics, chemistry, or certain quantifiable areas of biology.

In spite of these problems, various aspects of the environment can be determined by physical and chemical means, ranging from simple chemical identifications and physical measurements to the use of sophisticated mechanical apparatus. The development of biostatistics (statistics applied to the analysis of biological data), the elaboration of proper experimental design, and improved sampling methods now permit a quantified statistical approach to the study of ecology. Because of the extreme difficulties of controlling environmental variables in the field, studies involving the use of experimental design are largely confined to the laboratory and to controlled field experiments designed to test the effects of only one variable or several variables. The use of statistical procedures and computer models based on data obtained from the field provide insights into population interactions and ecosystem functions. Mathematical programming models are becoming increasingly important in applied ecology, especially in the management of natural resources and agricultural problems having an ecological basis.

Controlled environmental chambers enable experimenters to maintain plants and animals under known conditions of light, temperature, humidity, and day length so that the effects of each variable (or combination of variables) on the organism can be studied. Biotelemetry and other electronic tracking equipment, which allow the movements and behaviour of free-ranging organisms to be followed remotely, can provide rapid sampling of populations. Radioisotopes are used for tracing the pathways of nutrients through ecosystems, for determining the time and extent of transfer of energy and nutrients through the different components of the ecosystem, and for the determination of food chains. The use of laboratory microcosms—aquatic and soil micro-ecosystems, consisting of biotic and nonbiotic material from natural ecosystems, held under conditions similar to those found in the field—are useful in determining rates of nutrient cycling, ecosystem development, and other functional aspects of ecosystems. Microcosms enable the ecologist to duplicate experiments and to perform experimental manipulation on them.

Terrestrial Ecology

Terrestrial ecology is the study of land-based ecosystems, their populations and communities of plants, animals, and microbes, their interactions with the atmosphere and with streams and groundwater, and their role in the cycling of energy, water, and the major biogeochemical elements such as carbon and nitrogen. Research approaches include field measurement campaigns and experiments, laboratory analyses, analyzing satellite images to study variation across the landscape and through time, and computer modeling to test our understanding of how populations, communities, and ecosystems function at present and in response to environmental change. Humans affect terrestrial ecosystems through land and water management, pollution, and climate change.

Microbial Ecology

Microbial ecology is the study of the interactions of microorganisms with their environment, each other, and plant and animal species. It includes the study of symbioses, biogeochemical cycles and the interaction of microbes with anthropogenic effects such as pollution and climate change.

Microbes and Ecosystem Niches

Every ecosystem on Earth contains microorganisms that occupy unique niches based on their specific metabolic properties.

Microbial life is amazingly diverse and microorganisms quite literally cover the planet. In fact, it has been estimated that there are 100,000,000 times more microbial cells on the planet than there are stars in the observable universe. Microbes live in all parts of the biosphere where there is liquid water, including soil, hot springs, the ocean floor, acid lakes, deserts, geysers, rocks, and even the mammalian gut.

By virtue of their omnipresence, microbes impact the entire biosphere; indeed, microbial metabolic processes (including nitrogen fixation, methane metabolism, and sulfur metabolism) collectively control global biogeochemical cycling. The ability of microbes to contribute substantially to the function of every ecosystem is a reflection their tremendous biological diversity.

Microbes are vital to every ecosystem on Earth and are particularly important in zones where light cannot approach (that is, where photosynthesis cannot be the basic means to collect energy). Microorganisms participate in a host of fundamental ecological processes including production, decomposition, and fixation. They can also have additional indirect effects on the ecosystem through symbiotic relationships with other organisms. In addition, microbial processes can be co-opted for biodegradation or bioremediation of domestic, agricultural, and industrial wastes, making the study of microbial ecology particularly important for biotechnological and environmental applications.

Each species in an ecosystem is thought to occupy a separate, unique niche. The ecological niche of a microorganism describes how it responds to the distribution of resources and competing species, as well as the ways in which it alters those same factors in turn. In essence, the niche is a complex description of the ways in which a microbial species uses its environment.

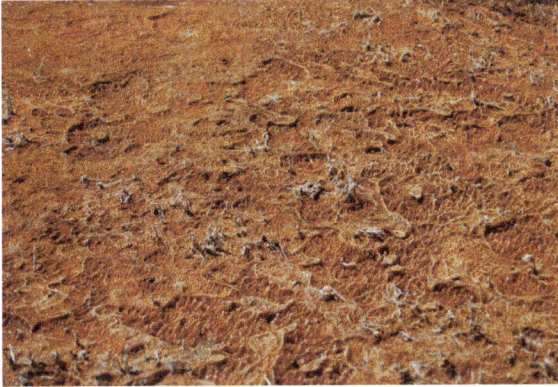

A Biofilm of Thermophilic Bacteria: Thermophiles, which thrive at relatively high temperatures, occupy a unique ecological niche. This image shows a colony of thermophilic bacteria at Mickey Hot Springs in Oregon.

The precise ecological niche of a microbe is primarily determined by the specific metabolic properties of that organism. For example, microbial organisms that can obtain energy from the oxidation of inorganic compounds (such as iron-reducing bacteria) will likely occupy a different niche from those that obtain energy from light (such as cyanobacteria). Even among photosynthetic bacteria, there are various species that contain different photosynthetic pigments (such as chlorophylls and carotenoids) that allow them to take advantage of different portions of the electromagnetic spectrum; therefore, even microbes with similar metabolic properties may inhabit unique niches.

Organization of Ecosystems

Microorganisms serve essential roles in the complex nutrient exchange system that defines an ecological community.

Although ecologists tend to regard ecosystems as basic structural units, it can be difficult (if not impossible) to formally define the boundaries of a given ecosystem. As such, ecosystems are better thought of as conceptual rather than actual geographical locations. Rarely are ecosystems isolated from one another; rather, they should be considered parts of a larger functioning whole that together comprise the biosphere ("the place on Earth's surface where life dwells").

Despite the fact that clear boundaries between ecosystems may be difficult to identify, the myriad interactions that take place within an ecological community can often be observed and defined. These interactions may be best described by detailing feeding connections (what eats what) among biota in an ecosystem, thereby linking the ecosystem into a unified system of exchange.

All life forms in an ecosystem can be broadly grouped into one of two categories (called trophic levels):

- Autotrophs, which produce organic matter (food) from inorganic substances; and

- Heterotrophs, which must feed on other organisms in order to obtain organic matter.

In general, trophic levels are used to describe the way in which a particular organism within an ecosystem gets its food. Using this description, we can restate and reorganize the categories above to define the three basic ways organisms acquire their food:

- Producers (autotrophs) do not usually eat other organisms but pull nutrients from the soil or the ocean and manufacture their own food using photosynthesis. In this way, it is the energy from the sun that usually powers the base of the food chain.

- Consumers (heterotrophs) cannot manufacture their own food and need to consume other organisms.

- Decomposers break down dead plant and animal material and wastes and release them into the ecosystem as energy and nutrients for recycling.

Within ecosystems, the biotic factors that comprise the categories above can be organized into a food chain in which autotrophic producers use materials and nutrients recycled by decomposers to make their own food; the producers are in turn eaten by heterotrophic consumers. In real world ecosystems, there are multiple food chains for most organisms (since most organisms eat more than one kind of food or are eaten by more than one type of predator). Additionally, the movement of mineral nutrients in the food chain is cyclic rather than linear. As a consequence, the intricate network of intersecting and overlapping food chains for an ecosystem is more commonly represented as a food web. A food web depicts a collection of heterotrophic consumers that network and cycle the flow of energy and nutrients from a productive base of self-feeding autotrophs.

Microorganisms play a vital role in every ecological community by serving both as producers and as decomposers. Although plants are the most common primary producers, autotrophic photosynthetic microbes (such as cyanobacteria and algae) can harness light energy to generate organic matter. Additionally, in zones where light cannot penetrate (and thus photosynthesis cannot be the basic means to produce energy), chemosynthetic microbes provide energy and carbon to the other organisms in the ecosystem. Other microbes are decomposers, with the ability to recycle nutrients from dead organic matter and other organisms' waste products. Decomposition is critical as most of the carbon and energy incorporated into plant tissues during photosynthesis remains uneaten when the plant tissue dies (and therefore must be broken down before it can be made available for recycling).

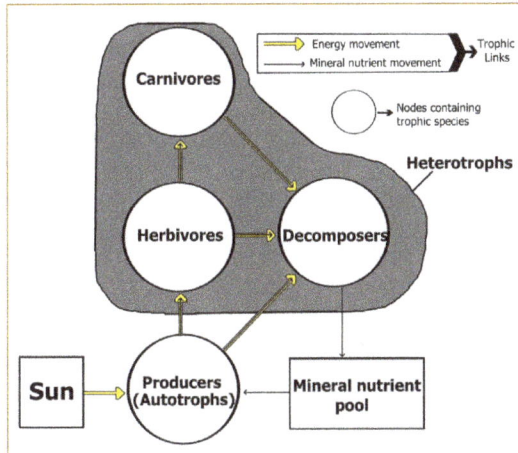

A simplified food web: This image shows a simplified food web model of energy and mineral nutrient movement in an ecosystem. Energy flow is unidirectional (noncyclic) and mineral nutrient movement is cyclic.

Role of Microbes in Biogeochemical Cycling

Microbes form the backbone of every ecological system by controlling global biogeochemical cycling of elements essential for life.

Microbial Role in Biogeochemical Cycling

Nutrients move through the ecosystem in biogeochemical cycles. A biogeochemical cycle is a pathway by which a chemical element (such as carbon or nitrogen) circulates through the biotic (living) and the abiotic (non-living) factors of an ecosystem. The elements that move through the factors of an ecosystem are not lost but are instead recycled or accumulated in places called reservoirs (or "sinks") where they can be held for a long period of time. Elements, chemical compounds, and other forms of matter are passed from one organism to another and from one part of the biosphere to another through these biogeochemical cycles.

Ecosystems have many biogeochemical cycles operating as a part of the system. A good example of a molecule that is cycled within an ecosystem is water, which is always recycled through the water cycle. Water undergoes evaporation, condensation, and then falls back to Earth as rain (or other forms of precipitation). This typifies the cycling that is observed for all of the principal elements of life.

Although biogeochemical cycles in a given ecosystem are coordinated by the full complement of living organisms and abiotic factors that make up that system, microorganisms play a primary role in regulating biogeochemical systems in virtually all of our planet's environments. This includes extreme environments such as acid lakes and hydrothermal vents, and even includes living systems such as the human gut. The key collective metabolic processes of microbes (including nitrogen fixation, carbon fixation, methane metabolism, and sulfur metabolism) effectively control global biogeochemical cycling. Incredibly, production by microbes is so immense that global biogeochemistry would likely not change even if eukaryotic life were totally absent.

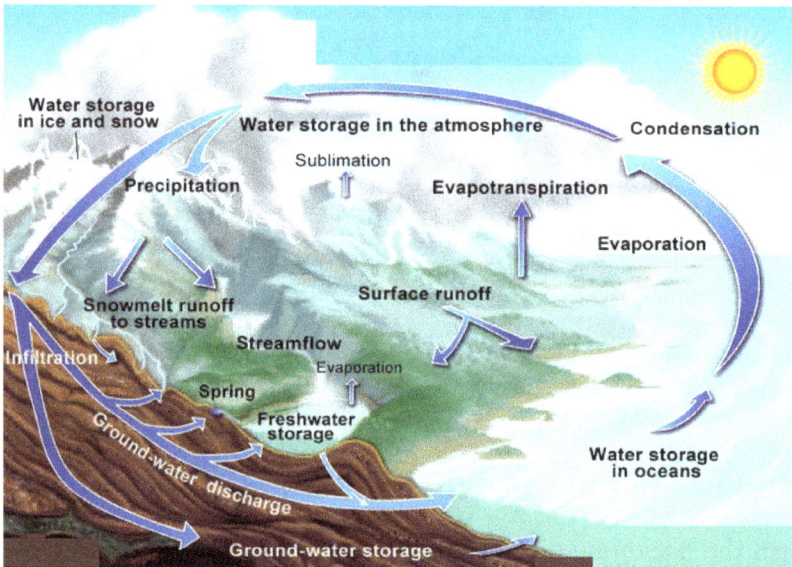

The Water Cycle: Water is recycled in an ecosystem through the water cycle.

Microbes comprise the backbone of every ecological system, particularly those in which there is no light (i.e. systems in which energy cannot be collected through photosynthesis).

The Carbon Cycle

Carbon is critical for life because it is the essential building block of all organic compounds. Plants and animals utilize carbon to produce carbohydrates, fats, and proteins, which can then be used to build their internal structures or to obtain energy.

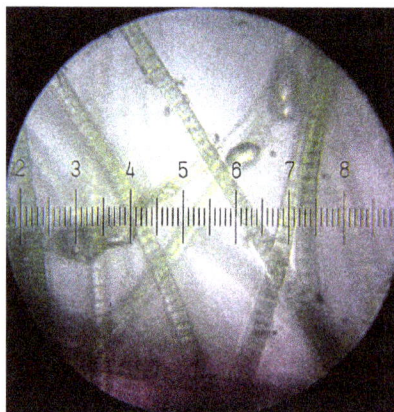

Cyanobacteria: Cyanobacteria, also known as blue-green bacteria, blue-green algae, and Cyanophyta, is a phylum of bacteria that obtain their energy through photosynthesis.

The Nitrogen Cycle

Nitrogen is essential for all forms of life because it is required for synthesis of the basic building blocks of life (e.g., DNA, RNA, and amino acids). The Earth's atmosphere is primarily composed of nitrogen, but atmospheric nitrogen (N_2) is relatively unusable for biological organisms. Consequently, chemical processing of nitrogen (or nitrogen fixation) is necessary to convert gaseous nitrogen into forms that living organisms can use. Almost all of the nitrogen fixation that occurs on the planet is carried out by bacteria that have the enzyme nitrogenase, which combines N_2 with hydrogen to produce a useful form of nitrogen (such as ammonia). Thus, microorganisms are absolutely essential for plant and animal life forms, which cannot fix nitrogen on their own.

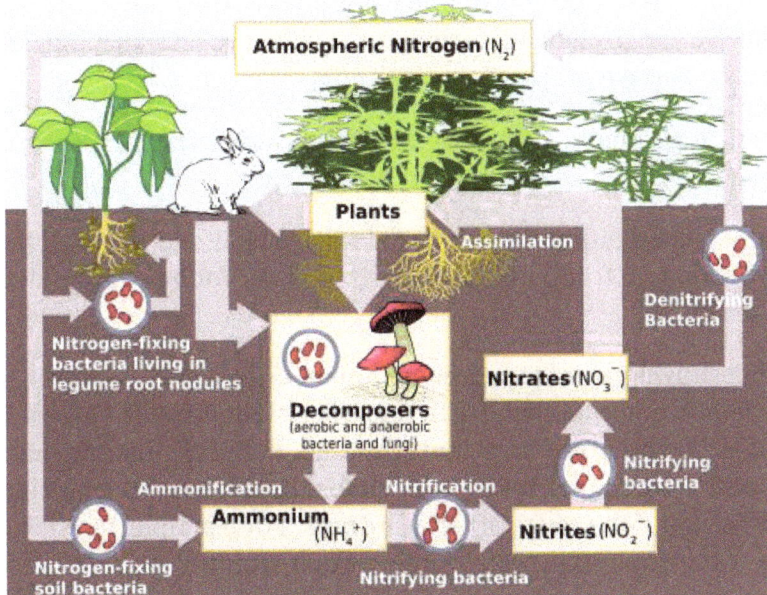

The Role of Microbes in the Nitrogen Cycle: The processing of nitrogen into a biologically useful form requires the activity of microorganisms.

Microbial Environments and Microenvironments

The extraordinary biological diversity among microbes reflects their ability to occupy every habitable environment on the planet.

Microorganisms are found on practically every habitable square inch of the planet. They live and thrive in all parts of the biosphere where there is liquid water, including hostile environments such as the poles, deserts, geysers, rocks, and the deep sea. Additionally, while microbes are often free-living, many have intimate symbiotic relationships with other larger organisms. Clearly, microbes have adapted to extreme and intolerant conditions, and it is this adaptation that has yielded tremendous biological diversity among microorganisms.

Like all extant organisms, microbes have evolved to thrive within a given environmental context. Microorganisms are ubiquitous despite the fact that the planet is host to extraordinarily diverse environments. Therefore, microbes have adapted to fill every ecological niche on the planet. For example, extremophilic species have been found that can tolerate the following environmental extremes:

- Temperatures as high as 130 °C (266 °F) and as low as −17 °C (1 °F).

- Highly alkaline (pH 0) and highly acidic (pH 11.5) environments.

- Extremely saline environments (including those in which the salt concentration is saturating).

- Extremely high (1,000-2,000 atm) and low (0 atm) pressures (some bacteria can survive for prolonged periods in a pressure-less vacuum, meaning they might even survive in space).

- High ionizing radiation (up to 15,000 Gy; as a reference, a mere 5 Gy would kill a human).

These evolutionary adaptations have allowed microbial life to extend into much of the Earth's atmosphere, crust, and hydrosphere (the water found over, under, and on the surface of a planet).

In addition to occupying a unique niche within an ecosystem, microbes are potentially sensitive to subtle environmental differences between adjacent areas. These differences define so-called microenvironments (or microhabitats) that can be distinguished from the immediate surroundings by such factors as the amount of incident light, the degree of moisture, and the range of temperatures. For example, the side of a tree that is shaded from sunlight is a microenvironment that typically supports a somewhat different community of microorganisms than would be found on the side that receives regular light. Microbes, therefore, are not only adapted to their habitat, but also to the immediate environment, thus promoting increased diversity among microbial species within an ecosystem.

Forest Ecology

Forest ecology is the study of forest ecosystems. Forests are ecosystems in which the major ecological characteristics reflect the dominance of ecosystem conditions and processes by trees. Ecosystems are ecological systems that have the attributes of structure, function, interaction of the component parts, complexity (that reflects the structure, function and interactions) and change over time. An ecosystem can be of almost any physical size as long as it exhibits these key characteristics, from a single plant growing in soil, to the entire world ecosystem.

The key structural components of forest ecosystems are plants, animals, microbes, soils and the atmosphere. Topography and microclimate are also important ecosystem features, but are not structural elements in the strict sense.

The key functional aspects of forest ecosystems are energy capture and biomass creation; nutrient cycling and the regulation of atmospheric and water chemistry; and important contributions to the regulation of the water cycle.

The interactions within an ecosystem involve all combinations of plant, animal and microbial interactions, interactions between organisms and the soil, and between the atmosphere and both the biotic community and the soil.

Complexity is an important attribute even though normally functioning forest ecosystems can exist at widely different levels of complexity. The importance of complexity lies in its implications for our ability to understand and predict, and therefore manage, forest ecosystems.

Forest ecosystems are continually changing. This change, initiated by external disturbance factors but largely determined by internal ecosystem processes, is vital for the maintenance of many aspects of biological diversity. In many types of forests it is essential for the long-term sustainability of the ecosystem.

"Forest stewardship" and "good, sustainable forestry" can only be defined in terms of society's desires and preferences with respect to stand and landscape-level forest conditions, functions and values. However, unless forestry is based on a respect for forest ecology and the ecological characteristics of forest ecosystems, it is very unlikely that society's long-term desires will be satisfied. Because of the long time scales of forestry, decisions about forest management must be founded on ecologically-based forecasts of ecosystem response, involving the use of ecosystem management simulation models.

Population Ecology

Population Ecology is the study of the processes that affect the distribution and abundance of animal and plant populations.

A population is a subset of individuals of one species that occupies a particular geographic area and, in sexually reproducing species, interbreeds. The geographic boundaries of a population are easy to establish for some species but more difficult for others. For example, plants or animals occupying islands have a geographic range defined by the perimeter of the island. In contrast, some species are dispersed across vast expanses, and the boundaries of local populations are more difficult to determine. A continuum exists from closed populations that are geographically isolated from, and lack exchange with, other populations of the same species to open populations that show varying degrees of connectedness.

Genetic Variation within Local Populations

In sexually reproducing species, each local population contains a distinct combination of genes. As a result, a species is a collection of populations that differ genetically from one another to a greater or lesser degree. These genetic differences manifest themselves as differences among populations in morphology, physiology, behaviour, and life histories; in other words, genetic characteristics (genotype) affect expressed, or observed, characteristics (phenotype). Natural selection initially operates on an individual organismal phenotypic level, favouring or discriminating against individuals based on their expressed characteristics. The gene pool (total aggregate of genes in a population at a certain time) is affected as organisms with phenotypes that are compatible with the environment are more likely to survive for longer periods, during which time they can reproduce more often and pass on more of their genes.

The amount of genetic variation within local populations varies tremendously, and much of the discipline of conservation biology is concerned with maintaining genetic diversity within and among populations of plants and animals. Some small isolated populations of asexual species often have little genetic variation among individuals, whereas large sexual populations often have great variation. Two major factors are responsible for this variety: mode of reproduction and population size.

Effects of Mode of Reproduction: Sexual and Asexual

In sexual populations, genes are recombined in each generation, and new genotypes may result. Offspring in most sexual species inherit half their genes from their mother and half from their father, and their genetic makeup is therefore different from either parent or any other individual in the population. In both sexually and asexually reproducing species, mutations are the single most important source of genetic variation. New favourable mutations that initially appear in separate individuals can be recombined in many ways over time within a sexual population.

In contrast, the offspring of an asexual individual are genetically identical to their parent. The only source of new gene combinations in asexual populations is mutation. Asexual populations accumulate genetic variation only at the rate at which their genes mutate. Favourable mutations arising in different asexual individuals have no way of recombining and eventually appearing together in any one individual, as they do in sexual populations.

Effects of Population Size

Over long periods of time, genetic variation is more easily sustained in large populations than in small populations. Through the effects of random genetic drift, a genetic trait can be lost from a small population relatively quickly. For example, many populations have two or more forms of a gene, which are called alleles. Depending on which allele an individual has inherited, a certain phenotype will be produced. If populations remain small for many generations, they may lose all but one form of each gene by chance alone.

This loss of alleles happens from sampling error. As individuals mate, they exchange genes. Imagine that initially half of the population has one form of a particular gene, and the other half of the population has another form of the gene. By chance, in a small population the exchange of genes could result in all individuals of the next generation having the same allele. The only way for this population to contain a variation of this gene again is through mutation of the gene or immigration of individuals from another population.

Minimizing the loss of genetic variation in small populations is one of the major problems faced by conservation biologists. Environments constantly change, and natural selection continually sorts through the genetic variation found within each population, favouring those individuals with phenotypes best suited for the current environment. Natural selection, therefore, continually works to reduce genetic variation within populations, but populations risk extinction without the genetic variation that allows populations to respond evolutionarily to changes in the physical environment, diseases, predators, and competitors.

Population Density and Growth

Life Histories and the Structure of Populations

An organism's life history is the sequence of events related to survival and reproduction that occur from birth through death. Populations from different parts of the geographic range that a species inhabits may exhibit marked variations in their life histories. The patterns of demographic variation seen within and among populations are referred to as the structure of populations. These variations include breeding frequency, the age at which reproduction begins, the number of times an individual reproduces during its lifetime, the number of offspring produced at each reproductive episode (clutch or litter size), the ratio of male to female offspring produced, and whether reproduction is sexual or asexual. These differences in life history characteristics can have profound effects on the reproductive success of individuals and the dynamics, ecology, and evolution of populations.

Of the many differences in life history that occur among populations, age at the time of first reproduction is one of the most important for understanding the dynamics and evolution of a population. All else being equal, natural selection will favour, within species, individuals that reproduce earlier than other individuals in the population, because by reproducing earlier an individual's genes enter the gene pool (the sum of a population's genetic material at a given time) sooner than those of other individuals that were born at the same time but have not reproduced. Nonetheless, the "all else being equal" qualification is an important one because delayed reproductive strategies that ensure larger and more-robust offspring may be selected for in some species of long-lived organisms. Precocial development (unusually early maturation) to reproduction may be favoured, however, if the genes of early reproducers begin to spread throughout the population. Individuals whose genetic makeup allows them to reproduce earlier in life will come to dominate a population if there is no counterbalancing advantage to those individuals that delay reproduction until later in life.

Not all populations, however, are made up of individuals that reproduce very early in life. In the course of a lifetime, an individual must devote energy and resources to physiological demands other than reproduction. This is referred to as the cost of reproduction. To reproduce successfully, a plant first may have to grow to a certain height and outcompete its neighbours, and an animal may have to devote energy to growth so that it can reach a size at which it can fend off predators and successfully compete for mates. In many populations, individuals that delay reproduction have a better chance of surviving and leaving offspring than those that attempt to reproduce early. The opposing demands of growth, defense, and reproduction are balanced within the constraints of different environments to produce populations that have a diverse range of life history strategies.

Populations often can be divided into one of two extreme types based on their life history strategy. Some populations, called r-selected, are considered opportunistic because their reproductive behaviour involves a high intrinsic rate of growth (r)—individuals give birth once at an early age to many offspring. Populations that exhibit this strategy often have been shaped by an extremely variable and uncertain environment. Because mortality occurs randomly in this setting, quantity of progeny rather than quality of care serves the species better. In another strategy, called K-selected, populations tend to remain near the carrying capacity (K), the maximum number of individuals that the environment can sustain. Individuals in a K-selected population give birth at a later age to fewer offspring. This equilibrial life history is exhibited in more stable environments where reproductive success depends more on the fitness of the offspring than on their numbers.

K-selected species.

Adult and young African savanna elephants (Loxodonta africana) crossing a stream. Elephants are classic examples of K-selected species—that is, species characterized by relatively stable populations. Such species produce a few large young instead of many small young.

Life Tables and the Rate of Population Growth

Differences in life history strategies, which include an organism's allocation of its time and resources to reproduction and care of offspring, greatly affect population dynamics. As stated above, populations in which individuals reproduce at an early age have the potential to grow much faster than populations in which individuals reproduce later. The effect of the age of first reproduction on population growth can be seen in the life tables for a particular species. Life tables were originally developed by insurance companies to provide a means of determining how long a person of a particular age could be expected to live. They are used not only by demographers of human populations but also by plant, animal, and microbial ecologists to make projections about the life expectancies of nonhuman populations, as well as the effects of variation on demography and population growth. The number of individuals in a closed population (a population in which neither immigration nor emigration occurs) is governed by the rates of birth (natality), growth, reproduction, and death (mortality). Life tables are designed to evaluate how these rates influence the overall growth rate of a population.

Survivorship Curves

Life tables follow the fate of a group of individuals all born within the same population in the same year. Of this group, or cohort, only a certain number of individuals will reach each age, and there is an age above which no individuals ever survive. Plotting the number of those members of the group that are still alive at each age results in a survivorship curve for the population. Survivorship curves are usually displayed on a semilogarithmic rather than an arithmetic scale.

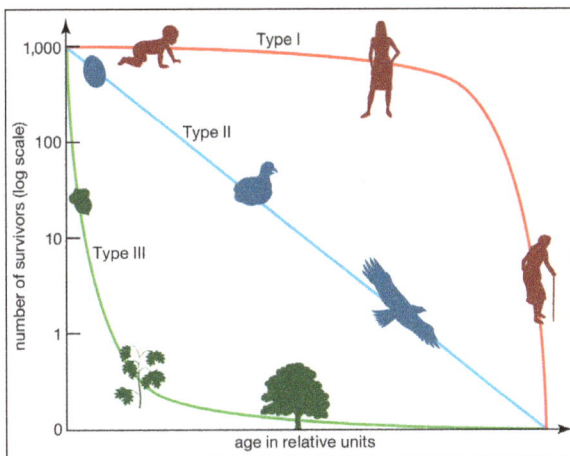

Survivorship curve.

Type I, II, and III survivorship curves. A survivorship curve is the graphic representation of the number of individuals in a population that can be expected to survive to any specific age.

There are three general types of survivorship curves. Species such as humans and other large mammals, which have fewer numbers of offspring but invest much time and energy in caring for their young (*K*-selected species), usually have a Type I survivorship curve. This relatively flat curve reflects low juvenile mortality, with most individuals living to old age. A constant probability of dying at any age, shown by the Type II survivorship curve, is evident as a straight line with a constant slope that decreases over time toward zero. Certain lizards, perching birds, and rodents exhibit this type of survivorship curve. In some species that produce many offspring but provide little care for them (*r*-selected species), mortality is greatest among the youngest individuals. The Type III survivorship curve indicative of this life history is initially very steep, which is reflective of very high mortality among the young, but flattens out as those individuals who reach maturity survive for a relatively longer time; it is exhibited by animals such as many insects or shellfish. Many populations have survivorship patterns that are more complex than, or fall in between, these three idealized curves. For example, passerine birds (perching birds such as finches) commonly suffer high mortality during the first year of life and a lower, more constant rate of death in subsequent years.

Calculating Population Growth

Life tables also are used to study population growth. The average number of offspring left by a female at each age together with the proportion of individuals surviving to each age can be used to evaluate the rate at which the size of the population changes over time. These rates are used by demographers and population ecologists to estimate population growth and to evaluate the effects of conservation efforts on endangered species.

Galapagos cactus finch.

It has such a high reproductive rate that the population can more than double in size each generation.

The average number of offspring that a female produces during her lifetime is called the net reproductive rate (R_o). If all females survived to the oldest possible age for that population, the net reproductive rate would simply be the sum of the average number of offspring produced by females at each age. In real populations, however, some females die at every age. The net reproductive rate for a set cohort is obtained by multiplying the proportion of females surviving to each age (l_x) by the average number of offspring produced at each age (m_x) and then adding the products from all the age groups: $R_o = \Sigma l_x m_x$. A net reproductive rate of 1.0 indicates that a population is neither increasing nor decreasing but replacing its numbers exactly. This rate indicates population stability. Any number below 1.0 indicates a decrease in population, while any number above indicates an increase. For example, the net reproductive rate for the Galapagos cactus finch (*Geospiza scandens*) is 2.101, which means that the population can more than double its size each generation.

age class** (x)	probability of surviving to age x (l_x)	average number of fledgling daughters (m_x)	product of survival and reproduction ($\Sigma l_x m_x$)
0	1.0	0.0	0.0
1	0.512	0.364	0.186
2	0.279	0.187	0.052
3	0.279	1.438	0.401
4	0.209	0.833	0.174
5	0.209	0.500	0.104
6	0.209	0.833	0.174
7	0.209	0.250	0.052
8	0.209	3.333	0.696
9	0.139	0.125	0.017
10	0.070	0.0	0.0
11	0.070	0.0	0.0
12	0.070	3.500	0.245
13	0	—	—
			R_o = 2.101

Net reproductive rate = $R_o = \Sigma l_x m_x$ = 2.101
Mean generation time = $T = (\Sigma x l_x m_x)/(R_o)$ = 6.08 years
Intrinsic rate of natural increase of the population = r = approximately $\ln R_o /T$ = 2.101/6.08 = 0.346
Life table for one Darwin finch, the Galapagos cactus finch (Geospiza scandens)*
*The values are for the cohort of females born in 1975.
**Designated in years.

The other value needed to calculate the rate at which the population can grow is the mean generation time (T). Generation time is the average interval between the birth of an individual and the birth of its offspring. To determine the mean generation time of a population, the age of the individuals (x) is multiplied by the proportion of females surviving to that age (lx) and the average number of offspring left by females at that age (m_x). This calculation is performed for each age group, and the values are added together and divided by the net reproductive rate (R_o) to yield the result For example, the mean generation time of the Galapagos cactus finch is 6.08 years:

$$T = \frac{\sum xl_x m_x}{R_o}$$

Another value is used by population biologists to calculate the rate of increase in populations that reproduce within discrete time intervals and possess generations that do not overlap. This is known as the intrinsic rate of natural increase (r), or the Malthusian parameter. Very simply, this rate can be understood as the number of births minus the number of deaths per generation time—in other words, the reproduction rate less the death rate. To derive this value using a life table, the natural logarithm of the net reproductive rate is divided by the mean generation time:

$$r\frac{\ln R_o}{T}$$

Values above zero indicate that the population is increasing; the higher the value, the faster the growth rate. The intrinsic rate of natural increase can be used to compare growth rates of populations of a species that have different generation times. Some human populations have higher intrinsic rates of natural increase partially because individuals in those groups begin reproducing earlier than those in other groups. Mice have higher intrinsic rates of natural increase than elephants because they reproduce at a much earlier age and have a much shorter mean generation time.

If a population has an intrinsic rate of natural increase of zero, then it is said to have a stable age distribution and neither grows nor declines in numbers. A growing population has more individuals in the lower age classes than does a stable population, and a declining population has more individuals in the older age classes than does a stable population. Many human populations are currently undergoing population increase, far exceeding a stable age distribution. Although the global human population has increased almost continuously throughout history, it has skyrocketed since the Industrial Revolution, primarily because of a drop in death rates. No other species has shown such sustained growth.

Species	Intrinsic rate of increase (r)
Elephant seal	0.091
Ring-necked pheasant	1.02
Field vole	3.18
Flour beetle	23
Water flea	69
Intrinsic rate of increase (r)* calculated for populations of species that differ greatly in their potential for the rate of population growth	
*Values above zero indicate that the population is increasing. The higher the value of r, the faster the intrinsic growth rate of the population.	

Regulation of Populations: Limits to Population Growth

Exponential and Geometric Population Growth

In an ideal environment, one that has no limiting factors, populations grow at a geometric rate or an exponential rate. Human populations, in which individuals live and reproduce for many years and in which reproduction is distributed throughout the year, grow exponentially. Exponential population growth can be determined by dividing the change in population size (ΔN) by the time interval (Δt) for a certain population size (N):

$$\frac{\Delta N}{\Delta t} = rN.$$

The growth curve of these populations is smooth and becomes increasingly steeper over time. The steepness of the curve depends on the intrinsic rate of natural increase for the population. Human population growth has been exponential since the beginning of the 20th century. Much concern exists about the impact this growth will have, not only on the environment but on humans as well. The World Bank projection for human population growth predicts that the human population will grow from 6.8 billion in 2010 to nearly 10 billion in 2050. That estimate could be offset by four population-control measures: (1) lower the rate of unwanted births, (2) lower the desired family size, (3) raise the average age at which women begin to bear children, and (4) reduce the number of births below the level that would replace current human populations (e.g., one child per woman).

Insects and plants that live for a single year and reproduce once before dying are examples of organisms whose growth is geometric. In these species a population grows as a series of increasingly steep steps rather than as a smooth curve.

Logistic Population Growth

The geometric or exponential growth of all populations is eventually curtailed by food availability, competition for other resources, predation, disease, or some other ecological factor. If growth is limited by resources such as food, the exponential growth of the population begins to slow as competition for those resources increases. The growth of the population eventually slows nearly to zero as the population reaches the carrying capacity (K) for the environment. The result is an S-shaped curve of population growth known as the logistic curve. It is determined by the equation:

$$\frac{\Delta N}{\Delta t} = rN\left(\frac{K-N}{K}\right).$$

Exponential versus logistic population growth

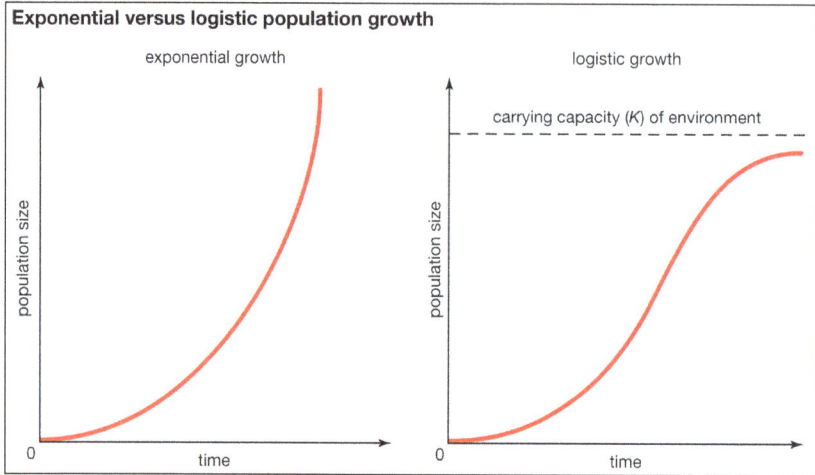

Carrying Capacity: Exponential versus Logistic Population Growth

In an ideal environment (one that has no limiting factors) populations grow at an exponential rate. The growth curve of these populations is smooth and becomes increasingly steep over time (left). However, for all populations, exponential growth is curtailed by factors such as limitations in food, competition for other resources, or disease. As competition increases and resources become increasingly scarce, populations reach the carrying capacity (K) of their environment, causing their growth rate to slow nearly to zero. This produces an S-shaped curve of population growth known as the logistic curve (right).

Population Fluctuation

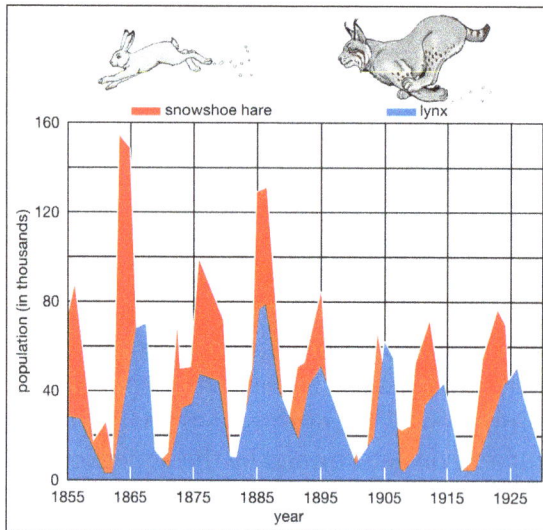

Cyclical fluctuations in the population density of the snowshoe hare and its effect on the population of its predator, the lynx. The graph is based on data derived from the records of the Hudson's Bay Company.

As stated above, populations rarely grow smoothly up to the carrying capacity and then remain there. Instead, fluctuations in population numbers, abundance, or density from one time step to the next are the norm. Population cycles make up a special type of population fluctuation, and the growth curves in population cycles are marked by distinct amplitudes and periods that set them apart from other population fluctuations. In a few species, such as snowshoe hares (*Lepus americanus*), lemmings, Canadian lynx (*Lynx canadensis*), and Arctic foxes (*Alopex lagopus*), populations show regular cycles of increase and decrease spanning a number of years. The causes of these fluctuations are still under debate by population ecologists, and no single cause may provide an explanation for every species. Most major hypotheses link regular fluctuations in population size to factors that are dependent on the density of the population, such as the availability of food or the activities of specialized predators, whose numbers track the abundance of their prey through population highs and lows.

Factors affecting Population Fluctuation

Population ecologists commonly divide the factors that affect the size of populations into density-dependent and density-independent factors. Density-independent factors, such as weather and climate, exert their influences on population size regardless of the population's density. In contrast, the effects of density-dependent factors intensify as the population increases in size. For example, some diseases spread faster in populations where individuals live in close proximity with one another than in those whose individuals live farther apart. Similarly, competition for food and other resources rises with density and affects an increasing proportion of the population. The dynamics of most populations are influenced by both density-dependent and density-independent factors, and the relative effects of the factors vary among populations. Density-independent factors are known as limiting factors, while density-dependent factors are sometimes called regulating factors because of their potential for maintaining population density within a narrow range of values.

Population Cycles

Because many factors influence population size, erratic variations in number are more common than regular cycles of fluctuation. Some populations undergo unpredictable and dramatic increases in numbers, sometimes temporarily increasing by 10 or 100 times over a few years, only to follow with a similarly rapid crash. For example, locusts in the arid parts of Africa multiply to such a level that their numbers can blacken the sky overhead; similar surges occurred in North America before the 20th century. The populations of some forest insects, such as the gypsy moths (*Lymantria dispar*) that were introduced to North America, rise extremely fast. As with species that fluctuate more regularly, the causes behind such sudden population increases are not fully known and are unlikely to have a single explanation that applies to all species.

The size of other populations varies within tighter limits. Some fluctuate close to their carrying capacity; others fluctuate below this level, held in check by various ecological factors, including predators and parasites. The tremendous expansion of many populations of weeds and pests that have been released into new environments in which their enemies are absent suggests that predators, grazers, and parasites all contribute to maintaining the small sizes of many populations.

Invasive prickly pear cactus (*Opuntia stricta*).

To control the explosive proliferation of these species, biological control programs have been instituted. With varying degrees of success, parasites or pathogens inimical to the foreign species have been introduced into the environment. The European rabbit (*Oryctolagus cuniculus*) was introduced into Australia in the 1800s, and its population grew unchecked, wreaking havoc on agricultural and pasture lands. The myxoma virus subsequently was released among the rabbit populations and greatly reduced them. Populations of the prickly pear cactus (*Opuntia*) in Australia and Africa grew unbounded until the moth borer (*Cactoblastis cactorum*) was introduced. However, many other similar attempts at biological control have failed, illustrating the difficulty in pinpointing the factors involved in population regulation.

Species Interactions and Population Growth

Interspecific Interactions

Community-level interactions are made up of the combined interactions between species within the biological community where the species coexist. The effects of one species upon another that derive from these interactions may take one of three forms: positive (+), negative (−), and neutral (0). Hence, interactions between any two species in any given biological community can take any of six forms:

- Mutualism (+, +), in which both species benefit from the interaction.

- Exploitation (+, −), in which one species benefits at the expense of the other.

- Commensalism (+, 0), in which one species benefits from the interaction while the other species neither benefits nor suffers.

- Interspecific competition (−, −), in which both species incur a cost of the interaction between them.

- Amensalism (−, 0), in which one species suffers while the other incurs no measurable cost of the interaction.

- Neutrality (0, 0), in which both species neither benefit nor suffer from the interaction.

Lotka-volterra Equations

The effects of species interactions on the population dynamics of the species involved can be predicted by a pair of linked equations that were developed independently during the 1920s by American mathematician and physical scientist Alfred J. Lotka and Italian physicist Vito Volterra. Today the Lotka-Volterra equations are often used to assess the potential benefits or demise of one species involved in competition with another species:

$$dN_1/dt = r_1 N_1 (1 − N_1/K_1 − \alpha_{1,2} N_2/K_2)$$

$$dN_2/dt = r_2 N_2 (1 − N_2/K_2 − \alpha_{2,1} N_1/K_1)$$

Here r = rate of increase, N = population size, and K = carrying capacity of any given species. In the first equation, the change in population size of species 1 over a specific period of time (dN_1/dt) is determined by its own population dynamics in the absence of species 2 ($r_1 N_1 [1 − N_1/K_1]$) as well as by its interaction with species 2 ($\alpha_{1,2} N_2/K_2$). As the formula implies, the effect of species 2 on species 1 ($\alpha_{1,2}$) in turn is determined by the population size and carrying capacity of species 2 (N_2 and K_2).

The possible outcomes of interactions between two species are predicted on the basis of the relative strengths of self-regulation versus the species interaction term. For instance, species 2 will drive species 1 to local extinction if the term $\alpha_{1,2} N_2/K_2$ exceeds the term $r_1 N_1 (1 − N_1/K_1)$—though the term $\alpha_{1,2} N_2/K_2$ will exert a decreasing influence over the growth rate of species 1 as $\alpha_{1,2} N_2/K_2$ diminishes. Consequently, the first equation represents the amount by which the growth rate of species 1 over a specific time period will be reduced by its interaction with species 2. In the second equation, the obverse applies to the dynamics of species 2.

In the case of interspecific competition, if the effects of both species on each other are approximately equivalent with respect to the strength of self-regulation in each species, the populations of both species may stabilize; however, one species may gradually exclude the other over time. The competitive exclusion scenario is dependent on the initial population size of each species. For instance, when the interspecific effects of each species upon the abundance of its competitor are approximately equal, the species with the higher initial abundance is likely to drive the species with a lower initial abundance to exclusion.

The basic equations given above, describing the dynamics deriving from an interaction between two competitors, have undergone several modifications. Chief among these modifications is the development of a subset of Lotka-Volterra equations that calculate the effects of interacting predator and prey populations. In their simplest forms, these modified equations bear a strong resemblance to the equations above, which are used to assess competition between two species:

$$dN_{prey}/dt = r_{prey} \times N_{prey}(1 - N_{prey}/K_{prey} - \alpha_{prey, pred} \times N_{pred}/K_{pred})$$

$$dN_{pred}/dt = r_{pred} \times N_{pred}(1 - N_{pred}/K_{pred} + \alpha_{pred, prey} \times N_{prey}/K_{prey})$$

Here the terms N_{pred} and K_{pred} denote the size of the predator population and its carrying capacity. Similarly, the population size and carrying capacity of the prey species are denoted by the terms N_{prey} and K_{prey}, respectively. The coefficient $\alpha_{prey, pred}$ represents the reduction in the growth rate of prey species due to its interaction with the predator, whereas $\alpha_{pred, prey}$ represents the increase in growth rate of the predator population due to its interaction with prey population.

Several additional modifications to the Lotka-Volterra equations are possible, many of which have focused on the incorporation of influences of spatial refugia (predator-free areas) from predation on prey dynamics.

Metapopulations

Although the dynamics and evolution of a single closed population are governed by its life history, populations of many species are not completely isolated and are connected by the movement of individuals (immigration and emigration) among them. Consequently, the dynamics and evolution of many populations are determined by both the population's life history and the patterns of movement of individuals between populations. Regional groups of interconnected populations are called metapopulations. These metapopulations are, in turn, connected to one another over broader geographic ranges. The mapped distribution of the perennial herb *Clematis fremontii* variety *Riehlii* in Missouri shows the metapopulation structure for this plant over an area of 1,129 square km (436 square miles). There is, therefore, a hierarchy of population structure from local populations to metapopulations to broader geographic groups of populations and eventually up to the worldwide collection of populations that constitute a species.

As local populations within a metapopulation fluctuate in size, they become vulnerable to extinction during periods when their numbers are low. Extinction of local populations is common in some species, and the regional persistence of such species is dependent on the existence of a metapopulation. Hence, elimination of much of the metapopulation structure of some species can increase the chance of regional extinction of species.

The structure of metapopulations varies among species. In some species one population may be particularly stable over time and act as the source of recruits into other, less stable populations. For example, populations of the checkerspot butterfly (*Euphydryas editha*) in California have a metapopulation structure consisting of a number of small satellite populations that surround a large source population on which they rely for new recruits. The satellite populations are too small and fluctuate too much to maintain themselves indefinitely. Elimination of the source population from this metapopulation would probably result in the eventual extinction of the smaller satellite populations.

Edith's checkerspot butterfly.

Edith's checkerspot butterfly (*Euphydryas editha*), male. This well-studied species inhabits a large portion of western North America and is known for being especially sensitive to annual variation in weather and climate.

In other species, metapopulations may have a shifting source. Any one local population may temporarily be the stable source population that provides recruits to the more unstable surrounding populations. As conditions change, the source population may become unstable, as when disease increases locally or the physical environment deteriorates. Meanwhile, conditions in another population that had previously been unstable might improve, allowing this population to provide recruits.

Overall, the population ecology and dynamics of all species is a complex result of their genetic structure, the life histories of the individuals, fluctuations in the carrying capacity of the environment, the relative influences of all the different kinds of density-dependent and density-independent factors that limit population growth, the spatial distribution of individuals, and the pattern of movement between populations that determines metapopulation structure. It is, therefore, not surprising that there are often great fluctuations in the numbers of individuals in local populations and that the long-term persistence of species may often require the conservation of many, rather than a few, populations.

Behavioral Ecology

Behavioral ecology examines the evolution of behaviors that allow animals to adapt to and thrive in their habitats.

There are two broad categories of behavior—learned and instinctive. Instinctive behavior is a pattern passed genetically from one generation to the next. A spider, for example, never needs to see another spider weave a web to know exactly how, where, and when to do it. This information is carried innately with the spider and allows it to carry out many of its life processes without ever having to think about them. The disadvantage to instinct is that is inflexible and does not allow the animal to change when the behavior is no longer appropriate. The armadillo's instinctive upward leap when threatened worked fine until the animal encountered a new environmental hazard—the automobile. Learned behavior, in contrast, is the result of experience accumulated and assimilated throughout a lifetime that allows the animal to adapt to unpredictable changes.

A behavioral ecologist studies patterns of behavior that fall somewhere between instinctive and learned. They include:

- Reflex: A rapid automatic response to a stimulus. Hedgehogs automatically curl into a ball when threatened.

- Conditioned reflex: An instinctive reflex that can be trained to occur under different conditions. A racehorse will go faster when flicked with a whip because it associates the whip with its traditional predator, a large cat, clawing at its back.

- Migration: A seasonal movement to a more favorable summer or winter environment. One of the most phenomenal migrations is that of the monarch butterfly, which spans thousands of miles and two generations. The young are genetically programmed to return to the fields their parents left.

- Hibernation and estivation: A state of torpor, or lowered metabolic rate resembling sleep, entered into by some animals in order to survive severely cold winters or hot, dry summers.

- Imprinting: Memorization by a young animal of the shape, sound, or smell of their parents or birthplace during a very brief period following birth. If the parent is absent, the baby will imprint on the first object it senses, giving rise to the sight of ducklings that think humans are their parents or kittens that have imprinted on dogs.

- Courtship: The special signals and complicated rituals that allow male-female bonds to occur for mating purposes. These behaviors assure the intentions and, consequently, the safety of both partners, who might attack or devour an approaching mate if the signals are unclear.

- Mimicry: The evolution of a harmless animal to look or behave like a dangerous animal. The viceroy butterfly mimics the coloration of the poisonous monarch, which most birds are genetically programmed to avoid.

- Preadaptation: A mixture of instinctive and learned behavior. Purple martins who once nested on cliffs have learned to use human-built structures to extend their ranges.

Applied Ecology

Applied ecology aims to relate ecological concepts, theories, principles, models, and methods to the solving of environmental problems, including the management of natural resources, such as land, energy, food or biodiversity.

Despite its somewhat restrictive name, applied ecology is more than simply the application of fundamental ecology. In a nutshell, ecological management requires prediction, and prediction requires theory. Applied ecology is a scientific field that studies how concepts, theories, models, or methods of fundamental ecology can be applied to solve environmental problems. It strives to fi nd practical solutions to these problems by comparing plausible options and determining, in the widest sense, the best management options.

One particular feature of applied ecology is that it uses an ecological approach to help solve questions concerned with specific parts of the environment, i.e., it considers a whole system and aims to account for all its inputs, outputs, and connections. Of course, accounting for everything is no more possible in applied ecology than it is in fundamental ecology, but the ecosystem approach of applied ecology is both one of its characteristics and one of its strengths.

Indeed, one could view the overall objective of applied ecology as to maintain the focal system while altering either some of the elements we take from the system (i.e., ecosystem services or exploitable resources) or some of those we add to the system (i.e., exploitation regimes or conservation measures) through an educated management strategy. Since those two types of elements are not mutually independent, long-term management strategies are best aimed at optimizing rather than maximizing exploited items. This is more efficiently achieved through an adequate understanding of theoretical ecology, which generally considers all parts of the system rather than a limited set of its components.

The word "applied" implies, directly or indirectly, human use or management of the environment and of its resources, either to preserve or restore them or to exploit them. Humans influence the Earth at all levels: the atmosphere, the hydrosphere (oceans and fresh water), the lithosphere (soil, land, and habitat), and the biosphere. Understandably, questions related to human populations (notably its demography) fall within the scope of applied ecology, as most impacts on ecosystems are directly or indirectly anthropogenic.

Aspects of applied ecology can be separated into two broad study categories: the outputs and the inputs. The first contains all fields dealing with the use and management of the environment for its ecosystem services and exploitable resources. These can be very diverse and include energy (fossil fuel or renewable energies), water, or soil. They can also be biological resources—for their exploitation—from fish to forests, to pastures and farmland. They might also, on the contrary, be species we wish to control: agricultural pests and weeds, alien invasive species, pollutants, parasites, and diseases. Finally, they can be species and spaces we wish to protect or to restore.

The fields devoted to studying the outputs of applied ecology include agro-ecosystem management, rangeland management, wildlife management (including game), landscape use (including development planning of rural, woodland, urban, and peri-urban regions), disturbance management (including fires and floods), environmental engineering, environmental design, aquatic resources management (including fisheries), forest management, and so on. This category also includes the use of ecological knowledge to control unwanted species: biological invasions, management of pests and weeds (including biological control), and epidemiology.

The inputs to an applied ecology problem consist of any management strategies or human influences on the target ecosystem or its biodiversity. These include conservation biology, ecosystem restoration, protected area design and management, global change, ecotoxicology and environmental pollution, bio-monitoring and bioindicators of environmental quality and biodiversity, environmental policies, and economics. Of course, these outputs and inputs are intimately connected. For example, the management of alien invasive species is relevant to both natural resource management (e.g., agriculture) and the protection of biodiversity (conservation/restoration biology).

In addition to using fundamental ecology to help solve practical environmental problems, applied ecology also aspires to facilitate resolutions by nonecologists, through a privileged dialogue with specialists of agriculture, engineering, education, law, policy, public health, rural and urban planning, natural resources management, and other disciplines for which the environment is a central axiom. Indeed, some of these disciplines are so influential on environmental management that they are viewed as inextricably interlinked. For example, conservation biology should really be named "conservation sciences" because it encompasses fields that are not very biological, such as environmental law, economics, administration and policy, philosophy and ethics, resources management, psychology, sociology, biotechnologies, and more generally, applied mathematics, physics, and chemistry.

And obviously, as we are dealing with environment and ecosystems, everything is connected, all questions are interrelated, all disciplines are linked, and all answers are interwoven. Understanding a process through one field of applied ecology will allow advancing knowledge in other fields.

Link between Applied and Theoretical Ecology

Because theoretical ecology may be defined as the use of models (in the widest sense) to explain patterns, suggest experiments, or make predictions in ecology, it is easy to see that relations between theoretical and applied ecology are bidirectional. Simply put, theory feeds application, but application also allows for the testing of theory. Indeed, applied ecological problems are used to assess and develop ecological theory. In this regard, these two aspects of ecology complement and stimulate each other.

Yet the links between theoretical and applied ecology can fray. Theoretical ecology operates within the bounds of plausibility. What is theoretically plausible, however, is not always ecologically realistic. And if it is not ecologically realistic, then theoretical ecology cannot be applied to interrogate real ecological situations. In other words, theoretical ecology cannot always be used in applied ecology. The links between theoretical and applied ecology range from spurious to robust and have been used with varying success in different fields. Fields that have benefited from theory include fisheries and forestry management and veterinary sciences and epidemiology (both human and nonhuman). However, some other fields of applied ecology have not (or not yet) benefited fully from ecological theories, concepts, and principles.

Ecological Succession

Ecological succession is a term developed by botanists to describe the change in structure of a community of different species, or ecosystem. The concept of ecological succession arose from a desire to understand how large and complex ecosystems like forests can exist in places known to be recently formed, such as volcanic islands. The different types of ecological succession exists during different phases of an ecosystem, and depend on how developed that ecosystem is. In the concept of ecological succession, ecosystems advance until they reach a climax community. In the climax community, all of the resources are efficiently used and the total mass of vegetation maxes out. Many forests that have not been disturbed in many years are examples of a climax community.

Types of Ecological Succession

Primary Succession

When the planet first formed, there was no soil. Hot magma and cold water make hard rocks, as seen by newly formed islands. Primary ecological succession is the process of small organisms and erosion breaking down these rocks into soil. Soil is then the foundation for higher forms of plant life. These higher forms can produce food for animals, which can then populate the area as well. Eventually, a barren landscape of rocks will progress through primary ecological succession to become a climax community. After years and years, the soil layer increases in thickness and harbors many nutrients and beneficial bacteria that are required to support advanced plant life. If this primary ecosystem is disturbed and wiped out, secondary succession can take place.

Secondary Succession

The first picture displays a climax community. As the frames progress, the community is destroyed by a fire. As long as the fire does not burn hot enough to destroy the soil and the organisms it harbors, secondary ecological succession will take place. Small plants will come back first. After they create a solid layer of vegetation, larger plants will be able to take root and become established. At first, small shrubs and trees will dominate. As the trees grow, they will begin to block the light from most of the ground, which will change the structure of the species below the canopy. Eventually, the ecosystem will arrive at a climax community, which may or may not be the similar to the original community. It all depends on which species colonize the area, and which seeds are able to germinate and thrive.

Cyclic Succession

Cyclic ecological succession happens within established communities and is merely a changing of the structure of the ecosystem on a cyclical basis. Some plants thrive at certain times of the year, and lay dormant the rest. Other organism, like cicadas, lay dormant for many years and emerge all at once, drastically changing the ecosystem.

Examples of Ecological Succession

Acadia National Park

Acadia National Park, in Maine, suffered a large wildfire in 1947 of over 10,000 acres. Being nearly 20% of the parks size, many were concerned that the park would be destroyed forever. Restoration efforts were left to nature however, and many think that the choice to not intervene paid off. While the first years were ugly, and only small plants could colonize the burnt soil, many years has led to a great amount of diversity in the tree species. While the trees before the fire were mostly evergreen trees, deciduous forests now dominate the landscape.

Coral Reef Ecological Succession

While ecological succession is a term coined by botanist, it also applies heavily to animal population that go through a disruption. Take, for instance, a coral reef. The coral reef as an ecosystem did not just pop into existence, but like many plant communities had to be formed over time through ecological succession. The primary ecological succession in a coral reef is the colonizing of rocks by small coral polyps. These polyps will grow and divide many times to create coral colonies. The shapes and shelter of the coral colonies eventually attract small fish and crustaceans that live in an around the coral. Smaller fish are food for larger fish, and eventually a fully functioning coral reef exists. The principles of ecological succession, while developed in context to plants, exists in all established ecosystems.

References

- Ecology, science: britannica.com, Retrieved 31 March, 2019
- Terrestrial-ecology-land-management, research-areas: umces.edu, Retrieved 14 July, 2019
- Microbial-ecology, boundless-microbiology: lumenlearning.com, Retrieved 17 May, 2019
- Population-ecology, science: britannica.com, Retrieved 19 April, 2019
- Behavioral-ecology, news-wires-white-papers-and-books, science: encyclopedia.com, Retrieved 5 February, 2019
- Ecological-succession: biologydictionary.net, Retrieved 26 July, 2019

Ecosystem and its Types

2

- **Components of Ecosystem**
- **Forest**
- **Desert Ecosystem**
- **Aquatic Ecosystem**
- **Grasslands**
- **Taiga**
- **Tundra**
- **Montane Ecosystems**

The community of living organisms along with the non-living components of their environment is called an ecosystem. The major types of ecosystem are desert ecosystem, aquatic ecosystem, grasslands and montane ecosystems. This chapter has been carefully written to provide an easy understanding of these types of ecosystem.

Ecosystem is the complex of living organisms, their physical environment, and all their interrelationships in a particular unit of space.

An ecosystem can be categorized into its abiotic constituents, including minerals, climate, soil, water, sunlight, and all other nonliving elements, and its biotic constituents, consisting of all its living members. Linking these constituents together are two major forces: the flow of energy through the ecosystem, and the cycling of nutrients within the ecosystem.

The fundamental source of energy in almost all ecosystems is radiant energy from the Sun. The energy of sunlight is used by the ecosystem's autotrophic, or self-sustaining, organisms. Consisting largely of green vegetation, these organisms are capable of photosynthesis—i.e., they can use the energy of sunlight to convert carbon dioxide and water into simple, energy-rich carbohydrates. The autotrophs use the energy stored within the simple carbohydrates to produce the more complex organic compounds, such as proteins, lipids, and starches, that maintain the organisms' life processes. The autotrophic segment of the ecosystem is commonly referred to as the producer level.

Organic matter generated by autotrophs directly or indirectly sustains heterotrophic organisms. Heterotrophs are the consumers of the ecosystem; they cannot make their own food. They use, rearrange, and ultimately decompose the complex organic materials built up by the autotrophs. All animals and fungi are heterotrophs, as are most bacteria and many other microorganisms.

Together, the autotrophs and heterotrophs form various trophic (feeding) levels in the ecosystem: the producer level, composed of those organisms that make their own food; the primary consumer level, composed of those organisms that feed on producers; the secondary consumer level, composed of those organisms that feed on primary consumers; and so on. The movement of organic matter and energy from the producer level through various consumer levels makes up a food chain. For example, a typical food chain in a grassland might be grass (producer) → mouse (primary consumer) → snake (secondary consumer) → hawk (tertiary consumer). Actually, in many cases the food chains of the ecosystem overlap and interconnect, forming what ecologists call a food web. The final link in all food chains is made up of decomposers, those heterotrophs that break down dead organisms and organic wastes. A food chain in which the primary consumer feeds on living plants is called a grazing pathway; that in which the primary consumer feeds on dead plant matter is known as a detritus pathway. Both pathways are important in accounting for the energy budget of the ecosystem.

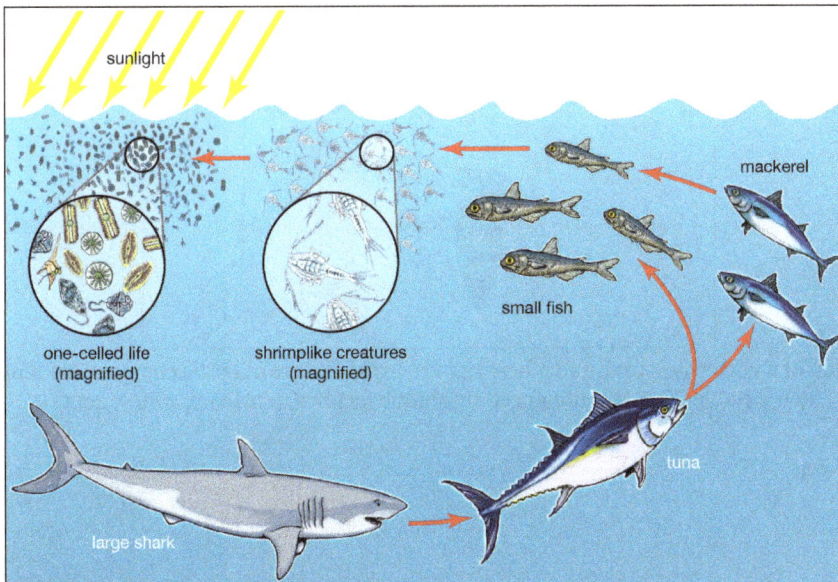

A food chain in the ocean begins with tiny one-celled organisms called diatoms. They make their own food from sunlight. Shrimplike creatures eat the diatoms. Small fish eat the shrimplike creatures, and bigger fish eat the small fish.

Components of Ecosystem

Biotic Components

Biotic components, or biotic factors, can be described as any living component that affects another organism or shapes the ecosystem. This includes both animals that consume other organisms within their ecosystem, and the organism that is being consumed. Biotic factors also include human influence, pathogens, and disease outbreaks. Each biotic factor needs the proper amount of energy and nutrition to function day to day.

Biotic components are typically sorted into three main categories:

- Producers, otherwise known as autotrophs, convert energy (through the process of photosynthesis) into food.

- Consumers, otherwise known as heterotrophs, depend upon producers (and occasionally other consumers) for food.

- Decomposers, otherwise known as detritivores, break down chemicals from producers and consumers (usually antibiotic) into simpler form which can be reused.

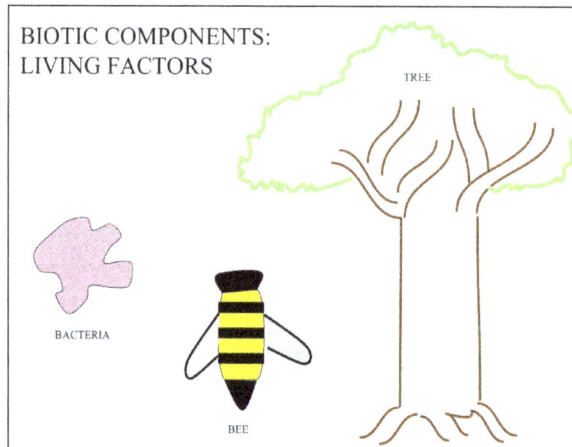

BIOTIC COMPONENTS: LIVING FACTORS

TREE

BACTERIA

BEE

Diagram of a bacterium, tree, and a bee that are living factors of biotic components found in an ecosystem that are influenced by abiotic factors (non-living components).

Influences

Species

Nearly all species are influenced by biotic factors in one way or another. If the number of predators was to increase, the entire food chain would be affected as any prey falling below that specified predator in the food chain will become prey. If the prey is not given enough time by the predator to repopulate, this could not only cause endangerment and extinction in the prey, but the predator as well. Contradicting a decrease in population size, if a particular species reproduces too rapidly, this will cause an increase in population size, thus affecting the environment around them.

Pathogens and Disease Outbreaks

When disease outbreaks occur, it can be detrimental to an ecosystem. When a disease hits, it will usually affect more than one species, thus causing a serious outbreak. This has the potential to set off a chain reaction thus, causing endangerment to a variety of species within that ecosystem.

Human Contact

Humans make the most sudden and long-term changes in an environment (e.g. pollution and waste). These changes either drive species out of their territory or force them to adapt to their new surroundings. These changes have the largest impact on an ecosystems population size, typically causing a serious decrease.

Biotic Components vs Abiotic Components

Biotic components are the living things that shape an ecosystem. Examples of biotic components include animals, plants, fungi, and bacteria. Abiotic components are non-living components that influence an ecosystem. Examples of abiotic factors are temperature, air currents, and minerals.

The factors mentioned above may either cause an increase or decrease in population size depending on the organism and ecosystem in question.

Abiotic Components

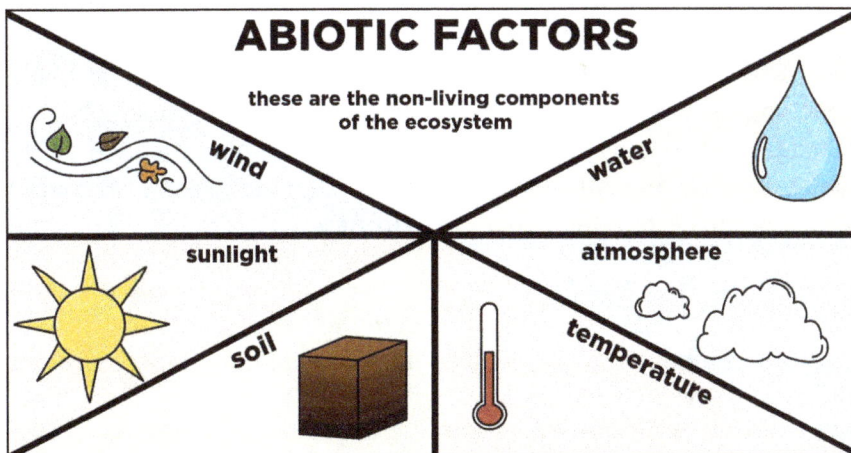

ABIOTIC FACTORS

these are the non-living components of the ecosystem

wind

water

sunlight

atmosphere

soil

temperature

Abiotic factors are non living components found in an ecosystem which influence living things (biotic factors).

In biology and ecology, abiotic components or abiotic factors are non-living chemical and physical parts of the environment that affect living organisms and the functioning of ecosystems. Abiotic factors and the phenomena associated with them underpin all biology.

Abiotic components include physical conditions and non-living resources that affect living organisms in terms of growth, maintenance, and reproduction. Resources are distinguished as substances or objects in the environment required by one organism and consumed or otherwise made unavailable for use by other organisms.

These are different settings on earth that are abiotic factors, which mean they are not living organisms, that contribute to the earth in many different ways.

Component degradation of a substance occurs by chemical or physical processes, e.g. hydrolysis. All non-living components of an ecosystem, such as atmospheric conditions and water resources, are called abiotic components.

In biology, abiotic factors can include water, light, radiation, temperature, humidity, atmosphere, acidity, and soil. The macroscopic climate often influences each of the above. Pressure and sound waves may also be considered in the context of marine or sub-terrestrial environments. Abiotic factors in ocean environments also include aerial exposure, substrate, water clarity, solar energy and tides. Consider the differences in the mechanics of C3, C4, and CAM plants in regulating the influx of carbon dioxide to the Calvin-Benson Cycle in relation to their abiotic stressors. C3 plants have no mechanisms to manage photorespiration, whereas C4 and CAM plants utilize a separate PEP Carboxylase enzyme to prevent photorespiration, thus increasing the yield of photosynthetic processes in certain high energy environments.

Many Archea require very high temperatures, pressures or unusual concentrations of chemical substances such as sulfur; this is due to their specialization into extreme conditions. In addition, fungi have also evolved to survive at the temperature, the humidity, and stability of their environment.

For example, there is a significant difference in access in both water and humidity between temperate rain forests and deserts. This difference in water availability causes a diversity in the organisms that survive in these areas. These differences in abiotic components alter the species present both by creating boundaries of what species can survive within the environment, as well as influencing competition between two species. Abiotic factors such as salinity can give one species a competitive advantage over another, creating pressures that lead to speciation and alteration of a species to and from generalist and specialist competitors.

Forest

A forest is a type of ecosystem in which there is high density of trees occupying a relatively large area of land. An ecosystem is an ecological unit consisting of a biotic community (an assemblage of plant, animal, and other living organisms) together with its abiotic environment (such as soil, rocks, water, temperature, slope of the land, etc). In the case of a forest, trees dominate the biotic landscape, although there are also other plants and animals. There are many types of forest, such as rainforests and temperate hardwood forest.

A dense growth of softwoods (a conifer forest) in the Sierra Nevada Range of Northern California.

There are many definitions of a forest, based on various criteria. These plant communities cover large areas of the globe and function as habitats for organisms, hydrologic flow modulators, and soil conservers, constituting one of the most important aspects of the Earth's biosphere.

Historically, "forest" meant an uncultivated area legally set aside for hunting by feudal nobility, and these hunting forests were not necessarily wooded much, if at all. However, as hunting forests did often include considerable areas of woodland, the word forest eventually came to mean wooded land more generally.

Forests provide innumerable values to people, provide aspects that address both physical needs (shelter, food, etc.) as well as the internal nature of people (beauty, diversity of animals and plants, recreation, etc). Some of the values of forests can be converted to commercial value, with trees providing timber for construction and housing, paper, firewood, and so forth. The selling of timber allows an economic valuation to be established. However, for many of the benefits offered by forests, it would be difficult to compute the economic value. These benefits include climate control, ecological values, water retention, and aesthetic, peace of mind, and recreational values.

Distribution

Forests can be found in all regions capable of sustaining tree growth, at altitudes up to the tree line, except where natural fire frequency is too high, or where the environment has been impaired by natural processes or by human activities. The tree line or timberline is the edge of the habitat at which trees are capable of growing; beyond the tree line, they are unable to grow due to inappropriate environmental conditions.

A deciduous broadleaf (Beech) forest in Slovenia.

As a general rule, forests dominated by angiosperms (broadleaf forests) are more species-rich than those dominated by gymnosperms (conifer, montane, or needleleaf forests), although exceptions exist (for example, species-poor aspen and birch stands in northern latitudes).

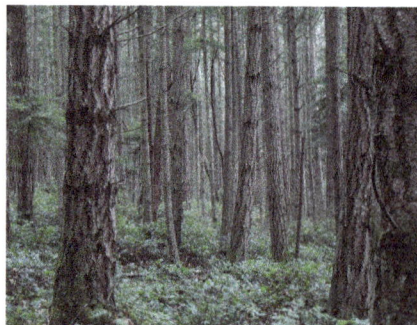

A forest on San Juan Island in Washington.

Forests sometimes contain many tree species within a small area (as in tropical rainforests and temperate deciduous forests), or relatively few species over large areas (e.g., taiga and arid montane coniferous forests). Forests are often home to many animal and plant species, and biomass per unit area is high compared to other vegetation communities. Much of this biomass occurs below-ground in the root systems and as partially decomposed plant detritus. The woody component of a forest contains lignin, which is relatively slow to decompose compared with other organic materials such as cellulose or carbohydrate.

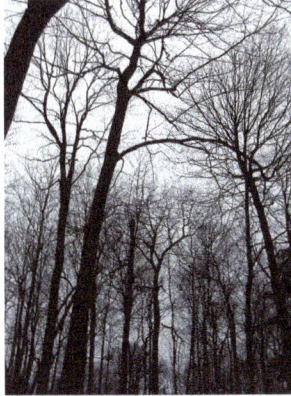

Maple and Oak (broadleaf, deciduous) forest in Wisconsin during winter.

Forests are differentiated from woodlands by the extent of canopy coverage: in a forest the branches and foliage of separate trees often meet or interlock, although there can be gaps of varying sizes within an area referred to as forest. A woodland has a more continuously open canopy, with trees spaced further apart, which allows more sunlight to penetrate to the ground between them.

Among the major forested biomes are:

- Rainforest (tropical and temperate);

- Taiga;

- Temperate hardwood forest;

- Tropical dry forest.

Forest Management and Forest Loss

Redwood tree in redwood forest, where many redwood trees are managed for preservation and longevity, rather than harvest for wood production.

The scientific study of forest species and their interaction with the environment is referred to as forest ecology, while the management of forests is often referred to as forestry. Forest management has changed considerably over the last few centuries, with rapid changes from the 1980s onwards culminating in a practice now referred to as sustainable forest management. Forest ecologists concentrate on forest patterns and processes, usually with the aim of elucidating cause and effect relationships. Foresters who practice sustainable forest management focus on the integration of ecological, social, and economic values, often in consultation with local communities and other stakeholders.

Anthropogenic factors that can affect forests include logging, human-caused forest fires, acid rain, and introduced species, among other things. For example, the gypsy moth, when introduced into North America from Eurasia, defoliated huge amounts of forest: Over 1 million acres (4,000 km^2) of forest each year were defoliated between 1980 and 1991, and in 1981 almost 13 million acres were defoliated (52,200 km^2). There are also many natural factors that can also cause changes in forests over time including forest fires, insects, diseases, weather, competition between species, and so forth. In 1997, the World Resources Institute recorded that only 20 percent of the world's original forests remained in large intact tracts of undisturbed forest. More than 75 percent of these intact forests lay in three countries: the Boreal forests of Russia and Canada, and the rainforest of Brazil.

Canada has about 4,020,000 km^2 of forest land. More than 90 percent of forest land is publicly owned and about 50 percent of the total forest area is allocated for harvesting. These allocated areas are managed using the principles of sustainable forest management, which includes extensive consultation with local stakeholders.

In the United States, most forests have historically been affected by humans to some degree, though in recent years improved forestry practices has helped regulate or moderate large scale or severe impacts. However the United States Forest Service estimates that every year about 6,000 km^2 (1.5 million acres) of the nation's 3,000,000 km^2 (750 million acres) of forest land is lost to urban sprawl and development. It is expected that the South alone will lose 80,000 to 100,000 km^2 (20 to 25 million acres) to development. However, in many areas of the United States, the area of forest is stable or increasing, particularly in many northern states.

Classification

Forests can be classified in different ways and to different degrees of specificity. One such way is in terms of the "biome" in which they exist, combined with leaf longevity of the dominant species (whether they are evergreen or deciduous). Another distinction is whether the forest is composed predominantly of broadleaf trees, coniferous trees, or mixed.

- Boreal forests occupy the subarctic zone and are generally evergreen and coniferous.

- Temperate zones support both broadleaf deciduous forests (e.g., temperate deciduous forest) and evergreen coniferous forests (e.g., Temperate coniferous forests and Temperate rainforests). Warm temperate zones support broadleaf evergreen forests, including laurel forests.

- Tropical and subtropical forests include tropical and subtropical moist forests, tropical and subtropical dry forests, and tropical and subtropical coniferous forests.

Physiognomy classifies forests based on their overall physical structure or developmental stage (e.g. old growth vs. second growth).

Forests can also be classified more specifically based on the climate and the dominant tree species present, resulting in numerous different forest types (e.g., ponderosa pine/Douglas-fir forest).

UNEP-WCMC Forest Category Classification

A number of global forest classification systems have been proposed but none has gained universal acceptance. A forest category classification has been developed by the United Nations Environment Program (UNEP) and the World Conservation Monitoring Center (WCMC). The UNEP-WCMC forest category classification system is a simplification of other more complex systems (e.g. UNESCO's forest and woodland "subformations"). This system divides the world's forest into 26 major types, which reflect climatic zones as well as the principal types of trees.

Broad Categories

The 26 major types in the UNEP-WCMC system can be reclassified into 6 broader categories:

- Temperate needleleaf: Temperate needleleaf forests mostly occupy the higher latitude regions of the northern hemisphere, as well as high altitude zones and some warm temperate areas, especially on nutrient-poor or otherwise unfavorable soils. These forests are composed entirely, or nearly so, of coniferous species. In the Northern Hemisphere, pines ‹›Pinus", spruces "Picea", larches "Larix", silver firs "Abies", Douglas firs "Pseudotsuga", and hemlocks ‹›Tsuga", make up the canopy, but other taxa are also important. In the southern hemisphere, mostly coniferous trees, members of the Araucariaceae and Podocarpaceae, occur in mixtures with broadleaf species that are classed as broadleaf and mixed forests.

- Temperate broadleaf and mixed: Temperate broadleaf and mixed forests include a substantial component of trees in the Anthophyta. They are generally characteristic of the warmer temperate latitudes, but extend to cool temperate ones, particularly in the southern hemisphere. They include such forest types as the mixed deciduous forests of the United States and their counterparts in China and Japan, the broadleaf evergreen rain forests of Japan, Chile, and Tasmania, the sclerophyllous forests of Australia, the Mediterranean and California, and the southern beech Nothofagus forests of Chile and New Zealand.

- Tropical moist: Tropical moist forests include many different forest types. The best known and most extensive are the lowland evergreen broadleaf rainforests include, for example: the seasonally inundated varzea and igapó forests and the terra firme forests of the Amazon Basin; the peat forests and moist dipterocarp forests of Southeast Asia; and the high forests of the Congo Basin. The forests of tropical mountains are also included in this broad category, generally divided into upper and lower montane formations on the basis of their physiognomy, which varies with altitude. The montane forests include cloud forest, those forests at middle to high altitude, which derive a significant part of their water budget from cloud, and support a rich abundance of vascular and nonvascular epiphytes. Mangrove forests also fall within this broad category, as do most of the tropical coniferous forests of Central America.

- Tropical dry: Tropical dry forests are characteristic of areas in the tropics affected by seasonal drought. The seasonality of rainfall is usually reflected in the deciduousness of the forest canopy, with most trees being leafless for several months of the year. However, under some conditions, such as less fertile soils or less predictable drought regimes, the proportion of evergreen species increases and the forests are characterized as "sclerophyllous". Thorn forest, a dense forest of low stature with a high frequency of thorny or spiny species, is found where drought is prolonged, and especially where grazing animals are plentiful. On very poor soils, and especially where fire is a recurrent phenomenon, woody savannas develop.

- Sparse trees and parkland: Sparse trees and parkland are forests with open canopies of 10-30 percent crown cover. They occur principally in areas of transition from forested to non-forested landscapes. The two major zones in which these ecosystems occur are in the boreal region and in the seasonally dry tropics. At high latitudes, north of the main zone of boreal forest or taiga, growing conditions are not adequate to maintain a continuous closed forest cover, so tree cover is both sparse and discontinuous. This vegetation is variously called open taiga, open lichen woodland, and forest tundra. It is species-poor, has high bryophyte cover, and is frequently affected by fire.

- Forest Plantations: Forest plantations, generally intended for the production of timber and pulpwood increase the total area of forest worldwide. Commonly mono-specific and composed of introduced tree species, these ecosystems are not generally important as habitat for native biodiversity. However, they can be managed in ways that enhance their biodiversity protection functions and they are important providers of ecosystem services, such as maintaining nutrient capital, protecting watersheds and soil structure, as well as storing carbon. They may also play an important role in alleviating pressure on natural forests for timber and fuelwood production.

Temperate Forest

Temperate forest is a forest found between the tropical and boreal regions, located in the temperate zone. It is within the second largest biome on the planet, covering 25% of the world's forest area, only behind the boreal forest, which covers about 33%. These forests cover both hemispheres at latitudes ranging from 25 to 50 degrees, wrapping the planet in a belt similar to that of the boreal forest. Due to its large size spanning several continents, there are several main types: deciduous, coniferous, broadleaf and mixed forest and rainforest.

Climate

The climate of a temperate forest is highly variable depending on the location of the forest. For example, Los Angeles and Vancouver, Canada are both considered to be located in a temperate zone, however, Vancouver is located in a temperate rainforest, while Los Angeles is more subtropical. Temperate forests typically have winters that often reach below freezing, however even this is not always true. The East Coast deciduous forests retain their deciduous nature largely due to the excessive freezing days each winter, as the leaves often freeze over and are only designed to live for one season. Milder areas such as the southern coast of British Columbia where the average winter lows are above freezing often have evergreen rainforests.

Types of Temperate Forests

Deciduous

They are found in Europe, East Asia, North America and in some parts of South America. Deciduous forests are composed mainly of broadleaf trees, such as maple and oak, that shed all their leaves during one season. They are typically found in three middle-latitude regions with temperate climates characterized by a winter season and year-round precipitation: eastern North America, western Eurasia and northeastern Asia.

Coniferous

Coniferous forests are composed of needle-leaved evergreen trees, such as pine or fir. Evergreen forests are typically found in regions with moderate climates. Boreal forests, however, are an exception as they are found in subarctic regions.

Broadleaf and Mixed

As the name implies, conifers and broadleaf trees grow in the same area. The main trees found in these forests are the great redwood, oak, ash, maple, birch, beech, poplar, elm and pine. Hardwood evergreen trees which are widely spaced and are found in the Mediterranean region are olive, cork, oak and stone pine.

Rainforest

Temperate rainforests are the wettest of all the types, and are found only in very wet coastal areas. Trees here are all evergreens, and are typically covered with thick moss and underbrush. Adding to its rarity is that most of the temperate rainforests outside of protected areas have been cut down and no longer exist. Currently, complete temperate rainforests can only be found in select areas of the Pacific Northwest and parts of Chile and New Zealand. Small stands can be found in Great Britain and southern Australia.

Effect of Human Activity

Temperate forests are located in the middle latitudes where much of the planet's population is. Not only were these forests cut down to build cities (i.e. New York City and Seattle), they have also been "cut down long ago to make way for cultivation". This biome has been subject to mining, logging, hunting, pollution, deforestation and habitat loss.

Tropical Rain Forests

Tropical rainforests are rainforests that occur in areas of tropical rainforest climate in which there is no dry season – all months have an average precipitation of at least 60 cm – and may also be referred to as *lowland equatorial evergreen rainforest*. True rainforests are typically found between 10 degrees north and south of the equator; they are a sub-set of the tropical forest biome that occurs roughly within the 28 degree latitudes (in the equatorial zone between the Tropic of Cancer and Tropic of Capricorn). Within the World Wildlife Fund's biome classification, tropical rainforests are a type of tropical moist broadleaf forest (or tropical wet forest) that also includes the more extensive seasonal tropical forests.

An area of the Amazon rainforest. The tropical rainforests
contain the largest diversity of species on Earth.

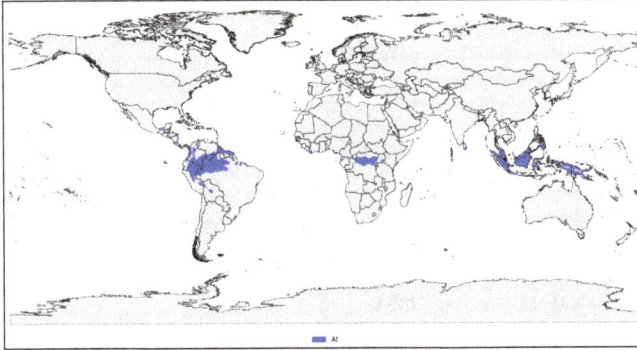

Tropical rainforest climate zones (Af).

Amazon River rain forest in Peru.

Tropical rainforests can be characterized in two words: hot and wet. Mean monthly temperatures exceed 18 °C (64 °F) during all months of the year. Average annual rainfall is no less than 1,680 mm (66 in) and can exceed 10 m (390 in) although it typically lies between 1,750 mm (69 in) and 3,000 mm (120 in). This high level of precipitation often results in poor soils due to leaching of soluble nutrients in the ground.

Tropical rainforests exhibit high levels of biodiversity. Around 40% to 75% of all biotic species are indigenous to the rainforests. Rainforests are home to half of all the living animal and plant species on the planet. Two-thirds of all flowering plants can be found in rainforests. A single hectare of rainforest may contain 42,000 different species of insect, up to 807 trees of 313 species and 1,500 species of higher plants. Tropical rainforests have been called the "world's largest pharmacy", because over one quarter of natural medicines have been discovered within them. It is likely that there may be many millions of species of plants, insects and microorganisms still undiscovered in tropical rainforests.

Tropical rainforests are among the most threatened ecosystems globally due to large-scale fragmentation as a result of human activity. Habitat fragmentation caused by geological processes such as volcanism and climate change occurred in the past, and have been identified as important drivers of speciation. However, fast human driven habitat destruction is suspected to be one of the major causes of species extinction. Tropical rain forests have been subjected to heavy logging and agricultural clearance throughout the 20th century, and the area covered by rainforests around the world is rapidly shrinking.

Other Types of Tropical Forest

Several biomes may appear similar to, or merge via ecotones with, tropical rainforest:

Moist Seasonal Tropical Forest

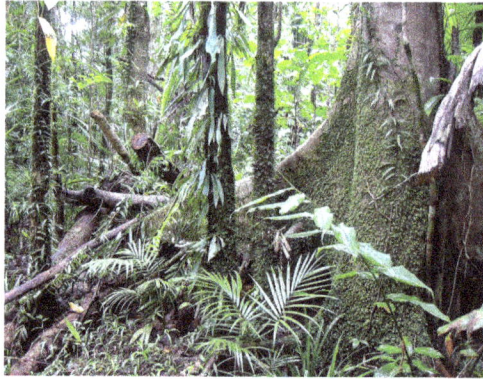

Daintree "rainforest" in Queensland is actually a seasonal tropical forest.

Moist seasonal tropical forests receive high overall rainfall with a warm summer wet season and a cooler winter dry season. Some trees in these forests drop some or all of their leaves during the winter dry season, thus they are sometimes called "tropical mixed forest". They are found in parts of South America, in Central America and around the Caribbean, in coastal West Africa, parts of the Indian subcontinent, and across much of Indochina.

Montane Rainforests

These are found in cooler-climate mountainous areas, becoming known as cloud forests at higher elevations. Depending on latitude, the lower limit of montane rainforests on large mountains is generally between 1500 and 2500 m while the upper limit is usually from 2400 to 3300 m.

Flooded Rainforests

Tropical freshwater swamp forests, or "flooded forests", are found in Amazon basin (the Várzea) and elsewhere.

Forest Structure

Rainforests are divided into different strata, or layers, with vegetation organized into a vertical pattern from the top of the soil to the canopy. Each layer is a unique biotic community containing different plants and animals adapted for life in that particular strata. Only the emergent layer is unique to tropical rainforests, while the others are also found in temperate rainforests.

Forest Floor

Western lowland gorilla.

The forest floor, the bottom-most layer, receives only 2% of the sunlight. Only plants adapted to low light can grow in this region. Away from riverbanks, swamps and clearings, where dense undergrowth is found, the forest floor is relatively clear of vegetation because of the low sunlight penetration. This more open quality permits the easy movement of larger animals such as: ungulates like the okapi (*Okapia johnstoni*), tapir (*Tapirus* sp.), Sumatran rhinoceros (*Dicerorhinus sumatrensis*), and apes like the western lowland gorilla (*Gorilla gorilla*), as well as many species of reptiles, amphibians, and insects. The forest floor also contains decaying plant and animal matter, which disappears quickly, because the warm, humid conditions promote rapid decay. Many forms of fungi growing here help decay the animal and plant waste.

Understory Layer

The understory layer lies between the canopy and the forest floor. The understory is home to a number of birds, small mammals, insects, reptiles, and predators. Examples include leopard (*Panthera pardus*), poison dart frogs (*Dendrobates* sp.), ring-tailed coati (*Nasua nasua*), boa constrictor (*Boa constrictor*), and many species of Coleoptera. The vegetation at this layer generally consists of shade-tolerant shrubs, herbs, small trees, and large woody vines which climb into the trees to capture sunlight. Only about 5% of sunlight breaches the canopy to arrive at the understory causing true understory plants to seldom grow to 3 m (10 feet). As an adaptation to these low light levels, understory plants have often evolved much larger leaves. Many seedlings that will grow to the canopy level are in the understory.

The canopy at the Forest Research Institute.

Canopy Layer

The canopy is the primary layer of the forest forming a roof over the two remaining layers. It contains the majority of the largest trees, typically 30–45 m in height. Tall, broad-leaved evergreen trees are the dominant plants. The densest areas of biodiversity are found in the forest canopy, as it often supports a rich flora of epiphytes, including orchids, bromeliads, mosses and lichens. These epiphytic plants attach to trunks and branches and obtain water and minerals from rain and debris that collects on the supporting plants. The fauna is similar to that found in the emergent layer, but more diverse. It is suggested that the total arthropod species richness of the tropical canopy might be as high as 20 million. Other species habituating this layer include many avian species such as the yellow-casqued wattled hornbill (*Ceratogymna elata*), collared sunbird (*Anthreptes collaris*), grey parrot (*Psitacus erithacus*), keel-billed toucan (*Ramphastos sulfuratus*), scarlet macaw (*Ara macao*) as well as other animals like the spider monkey (*Ateles* sp.), African giant swallowtail (*Papilio antimachus*), three-toed sloth (*Bradypus tridactylus*), kinkajou (*Potos flavus*), and tamandua (*Tamandua tetradactyla*).

Emergent Layer

The emergent layer contains a small number of very large trees, called *emergents*, which grow above the general canopy, reaching heights of 45–55 m, although on occasion a few species will grow to 70–80 m tall. Some examples of emergents include: *Balizia elegans*, *Dipteryx panamensis*, *Hieronyma alchorneoides*, *Hymenolobium mesoamericanum*, *Lecythis ampla* and *Terminalia oblonga*. These trees need to be able to withstand the hot temperatures and strong winds that occur above the canopy in some areas. Several unique faunal species inhabit this layer such as the crowned eagle (*Stephanoaetus coronatus*), the king colobus (*Colobus polykomos*), and the large flying fox (*Pteropus vampyrus*).

However, stratification is not always clear. Rainforests are dynamic and many changes affect the structure of the forest. Emergent or canopy trees collapse, for example, causing gaps to form. Openings in the forest canopy are widely recognized as important for the establishment and growth of rainforest trees. It is estimated that perhaps 75% of the tree species at La Selva Biological Station, Costa Rica are dependent on canopy opening for seed germination or for growth beyond sapling size, for example.

Ecology

Climates

Artificial tropical rainforest in Barcelona.

Tropical rainforests are located around and near the equator, therefore having what is called an equatorial climate characterized by three major climatic parameters: temperature, rainfall, and dry season intensity. Other parameters that affect tropical rainforests are carbon dioxide concentrations, solar radiation, and nitrogen availability. In general, climatic patterns consist of warm temperatures and high annual rainfall. However, the abundance of rainfall changes throughout the year creating distinct moist and dry seasons. Tropical forests are classified by the amount of rainfall received each year, which has allowed ecologists to define differences in these forests that look so similar in structure. According to Holdridge's classification of tropical ecosystems, true tropical rainforests have an annual rainfall greater than 2 m and annual temperature greater than 24 degrees Celsius, with a potential evapotranspiration ratio (PET) value of <0.25. However, most lowland tropical forests can be classified as tropical moist or wet forests, which differ in regards to rainfall. Tropical forest ecology- dynamics, composition, and function- are sensitive to changes in climate especially changes in rainfall.

Soils

Soil Types

Soil types are highly variable in the tropics and are the result of a combination of several variables such as climate, vegetation, topographic position, parent material, and soil age. Most tropical soils are characterized by significant leaching and poor nutrients, however there are some areas that contain fertile soils. Soils throughout the tropical rainforests fall into two classifications which include the ultisols and oxisols. Ultisols are known as well weathered, acidic red clay soils, deficient in major nutrients such as calcium and potassium. Similarly, oxisols are acidic, old, typically reddish, highly weathered and leached, however are well drained compared to ultisols. The clay content of ultisols is high, making it difficult for water to penetrate and flow through. The reddish color of both soils is the result of heavy heat and moisture forming oxides of iron and aluminium, which are insoluble in water and not taken up readily by plants.

Soil chemical and physical characteristics are strongly related to above ground productivity and forest structure and dynamics. The physical properties of soil control the tree turnover rates whereas chemical properties such as available nitrogen and phosphorus control forest growth rates. The soils of the eastern and central Amazon as well as the Southeast Asian Rainforest are old and mineral poor whereas the soils of the western Amazon (Ecuador and Peru) and volcanic areas of Costa Rica are young and mineral rich. Primary productivity or wood production is highest in western Amazon and lowest in eastern Amazon which contains heavily weathered soils classified as oxisols. Additionally, Amazonian soils are greatly weathered, making them devoid of minerals like phosphorus, potassium, calcium, and magnesium, which come from rock sources. However, not all tropical rainforests occur on nutrient poor soils, but on nutrient rich floodplains and volcanic soils located in the Andean foothills, and volcanic areas of Southeast Asia, Africa, and Central America.

Oxisols, infertile, deeply weathered and severely leached, have developed on the ancient Gondwanan shields. Rapid bacterial decay prevents the accumulation of humus. The concentration of iron and aluminium oxides by the laterization process gives the oxisols a bright red color and sometimes produces minable deposits (e.g., bauxite). On younger substrates, especially of volcanic origin, tropical soils may be quite fertile.

Nutrient Recycling

This high rate of decomposition is the result of phosphorus levels in the soils, precipitation, high temperatures and the extensive microorganism communities. In addition to the bacteria and other microorganisms, there are an abundance of other decomposers such as fungi and termites that aid in the process as well. Nutrient recycling is important because below ground resource availability controls the above ground biomass and community structure of tropical rainforests. These soils are typically phosphorus limited, which inhibits net primary productivity or the uptake of carbon. The soil contains microbial organisms such as bacteria, which break down leaf litter and other organic matter into inorganic forms of carbon usable by plants through a process called decomposition. During the decomposition process the microbial community is respiring, taking up oxygen and releasing carbon dioxide. The decomposition rate can be evaluated by measuring the uptake of oxygen. High temperatures and precipitation increase decomposition rate, which allows plant litter to rapidly decay in tropical regions, releasing nutrients that are immediately taken up by plants through surface or ground waters. The seasonal patterns in respiration are controlled by leaf litter fall and precipitation, the driving force moving the decomposable carbon from the litter to the soil. Respiration rates are highest early in the wet season because the recent dry season results in a large percentage of leaf litter and thus a higher percentage of organic matter being leached into the soil.

Buttress Roots

A common feature of many tropical rainforests is the distinct buttress roots of trees. Instead of penetrating to deeper soil layers, buttress roots create a widespread root network at the surface for more efficient uptake of nutrients in a very nutrient poor and competitive environment. Most of the nutrients within the soil of a tropical rainforest occur near the surface because of the rapid turnover time and decomposition of organisms and leaves. Because of this, the buttress roots occur at the surface so the trees can maximize uptake and actively compete with the rapid uptake of other trees. These roots also aid in water uptake and storage, increase surface area for gas exchange, and collect leaf litter for added nutrition. Additionally, these roots reduce soil erosion and maximize nutrient acquisition during heavy rains by diverting nutrient rich water flowing down the trunk into several smaller flows while also acting as a barrier to ground flow. Also, the large surface areas these roots create provide support and stability to rainforests trees, which commonly grow to significant heights. This added stability allows these trees to withstand the impacts of severe storms, thus reducing the occurrence of fallen trees.

Forest Succession

Succession is an ecological process that changes the biotic community structure over time towards a more stable, diverse community structure after an initial disturbance to the community. The initial disturbance is often a natural phenomenon or human caused event. Natural disturbances include hurricanes, volcanic eruptions, river movements or an event as small as a fallen tree that creates gaps in the forest. In tropical rainforests, these same natural disturbances have been well documented in the fossil record, and are credited with encouraging speciation and endemism. Human land use practices have lead to large scale deforestation. In many tropical countries such as Costa Rica these deforested lands have been abandoned and forests have been allowed to regenerate through ecological succession. These regenerating young successional forests are called secondary forests or second-growth forests.

Biodiversity and Speciation

Tropical rainforests exhibit a vast diversity in plant and animal species. The root for this remarkable speciation has been a query of scientists and ecologists for years. A number of theories have been developed for why and how the tropics can be so diverse.

Young orangutan at Bukit Lawang, Sumatra.

Interspecific Competition

Interspecific competition results from a high density of species with similar niches in the tropics and limited resources available. Species which lose the competition may either become extinct or find a new niche. Direct competition will often lead to one species dominating another by some advantage, ultimately driving it to extinction. Niche partitioning is the other option for a species. This is the separation and rationing of necessary resources by utilizing different habitats, food sources, cover or general behavioral differences. A species with similar food items but different feeding times is an example of niche partitioning.

Pliestocene Refugia

The theory of Pleistocene refugia was developed by Jürgen Haffer in 1969 with his article *Speciation of Amazonian Forest Birds*. Haffer proposed the explanation for speciation was the product of rainforest patches being separated by stretches of non forest vegetation during the last glacial period. He called these patches of rainforest areas refuges and within these patches allopatric speciation occurred. With the end of the glacial period and increase in atmospheric humidity, rainforest began to expand and the refuges reconnected. This theory has been the subject of debate. Scientists are still skeptical of whether or not this theory is legitimate. Genetic evidence suggests speciation had occurred in certain taxa 1–2 million years ago, preceding the Pleistocene.

Human Dimensions

Habitation

Tropical rainforests have harboured human life for many millennia, with many Indian tribes in South- and Central America, who belong to the Indigenous peoples of the

Americas, the Congo Pygmies in Central Africa, and several tribes in South-East Asia, like the Dayak people and the Penan people in Borneo. Food resources within the forest are extremely dispersed due to the high biological diversity and what food does exist is largely restricted to the canopy and requires considerable energy to obtain. Some groups of hunter-gatherers have exploited rainforest on a seasonal basis but dwelt primarily in adjacent savanna and open forest environments where food is much more abundant. Other people described as rainforest dwellers are hunter-gatherers who subsist in large part by trading high value forest products such as hides, feathers, and honey with agricultural people living outside the forest.

Indigenous Peoples

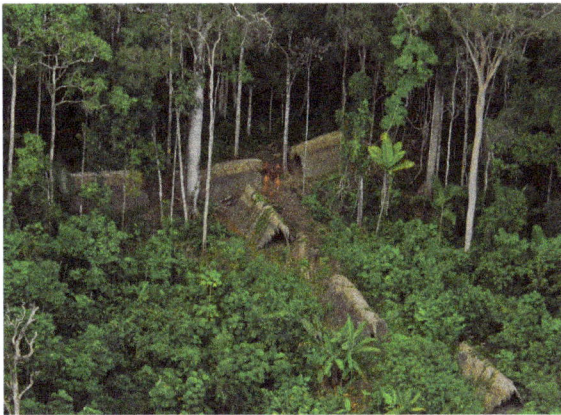

Members of an uncontacted tribe.

A variety of indigenous people live within the rainforest as hunter-gatherers, or subsist as part-time small scale farmers supplemented in large part by trading high-value forest products such as hides, feathers, and honey with agricultural people living outside the forest. Peoples have inhabited the rainforests for tens of thousands of years and have remained so elusive that only recently have some tribes been discovered. These indigenous peoples are greatly threatened by loggers in search for old-growth tropical hardwoods like Ipe, Cumaru and Wenge, and by farmers who are looking to expand their land, for cattle(meat), and soybeans, which are used to feed cattle in Europe and China. On 18 January 2007, FUNAI reported also that it had confirmed the presence of 67 different uncontacted tribes in Brazil, up from 40 in 2005. With this addition, Brazil has now overtaken the island of New Guinea as the country having the largest number of uncontacted tribes. The province of Irian Jaya or West Papua in the island of New Guinea is home to an estimated 44 uncontacted tribal groups.

The pygmy peoples are hunter-gatherer groups living in equatorial rainforests characterized by their short height (below one and a half meters, or 59 inches, on average). Amongst this group are the Efe, Aka, Twa, Baka, and Mbuti people of Central Africa. However, the term pygmy is considered pejorative so many tribes prefer not to be labeled as such.

Pygmy hunter-gatherers in the Congo Basin.

Some notable indigenous peoples of the Americas, or Amerindians, include the Huaorani, Yąnomamö, and Kayapo people of the Amazon. The traditional agricultural system practiced by tribes in the Amazon is based on swidden cultivation (also known as slash-and-burn or shifting cultivation) and is considered a relatively benign disturbance. In fact, when looking at the level of individual swidden plots a number of traditional farming practices are considered beneficial. For example, the use of shade trees and fallowing all help preserve soil organic matter, which is a critical factor in the maintenance of soil fertility in the deeply weathered and leached soils common in the Amazon.

There is a diversity of forest people in Asia, including the Lumad peoples of the Philippines and the Penan and Dayak people of Borneo. The Dayaks are a particularly interesting group as they are noted for their traditional headhunting culture. Fresh human heads were required to perform certain rituals such as the Iban "kenyalang" and the Kenyah "mamat". Pygmies who live in Southeast Asia are, amongst others, referred to as "Negrito".

Resources

Cultivated Foods and Spices

Yam, coffee, chocolate, banana, mango, papaya, macadamia, avocado, and sugarcane all originally came from tropical rainforest and are still mostly grown on plantations in regions that were formerly primary forest. In the mid-1980s and 1990s, 40 million tons of bananas were consumed worldwide each year, along with 13 million tons of mango. Central American coffee exports were worth US$3 billion in 1970. Much of the genetic variation used in evading the damage caused by new pests is still derived from resistant wild stock. Tropical forests have supplied 250 cultivated kinds of fruit, compared to only 20 for temperate forests. Forests in New Guinea alone contain 251 tree species with edible fruits, of which only 43 had been established as cultivated crops by 1985.

Ecosystem Services

In addition to extractive human uses, rain forests also have non-extractive uses that are frequently summarized as ecosystem services. Rain forests play an important role in maintaining biological diversity, sequestering and storing carbon, global climate regulation, disease control, and pollination. Half of the rainfall in the Amazon area is produced by the forests. The moisture from the forests is important to the rainfall in Brazil, Paraguay, Argentina Deforestation in the Amazon rainforest region was one of the main reason that cause the severe Drought of 2014-2015 in Brazil.

Conservation

Mining and Drilling

The Ok Tedi Mine in southwestern Papua New Guinea.

Deposits of precious metals (gold, silver, coltan) and fossil fuels (oil and natural gas) occur underneath rainforests globally. These resources are important to developing nations and their extraction is often given priority to encourage economic growth. Mining and drilling can require large amounts of land development, directly causing deforestation. In Ghana, a West African nation, deforestation from decades of mining activity left about 12% of the country's original rainforest intact.

Conversion to Agricultural Land

With the invention of agriculture, humans were able to clear sections of rainforest to produce crops, converting it to open farmland. Such people, however, obtain their food primarily from farm plots cleared from the forest and hunt and forage within the forest to supplement this. The issue arising is between the independent farmer providing for his family and the needs and wants of the globe as a whole. This issue has seen little improvement because no plan has been established for all parties to be aided.

Agriculture on formerly forested land is not without difficulties. Rainforest soils are often thin and leached of many minerals, and the heavy rainfall can quickly leach nutrients from area cleared for cultivation. People such as the Yanomamo of the Amazon, utilize slash-and-burn agriculture to overcome these limitations and enable them to push deep into what were previously rainforest environments. However, these are not rainforest dwellers, rather they are dwellers in cleared farmland that make forays into the rainforest. Up to 90% of the typical Yanamomo diet comes from farmed plants.

Some action has been taken by suggesting fallow periods of the land allowing secondary forest to grow and replenish the soil. Beneficial practices like soil restoration and conservation can benefit the small farmer and allow better production on smaller parcels of land.

Climate Change

The tropics take a major role in reducing atmospheric carbon dioxide. The tropics (most notably the Amazon rainforest) are called carbon sinks. As major carbon reducers and carbon and soil methane storages, their destruction contributes to increasing global energy trapping, atmospheric gases. Climate change has been significantly contributed to by the destruction of the rainforests. A simulation was performed in which all rainforest in Africa were removed. The simulation showed an increase in atmospheric temperature by 2.5 to 5 degrees Celsius.

Protection

Efforts to protect and conserve tropical rainforest habitats are diverse and widespread. Tropical rainforest conservation ranges from strict preservation of habitat to finding sustainable management techniques for people living in tropical rainforests. International policy has also introduced a market incentive program called Reducing Emissions from Deforestation and Forest Degradation (REDD) for companies and governments to outset their carbon emissions through financial investments into rainforest conservation.

Desert Ecosystem

A desert ecosystem is defined by interactions between organism populations, the climate in which they live, and any other non-living influences on the habitat. Deserts are arid regions which are generally associated with warm temperatures, however cold deserts also exist.

Regardless of the region, any desert is usually cold at night and receives very little rainfall. However, they do produce plants, which have adapted to such living conditions.

Several things make up a desert ecosystem. Among those are:

- Structure.

- Characteristics.

- Animals.

The ecosystem is dependent upon the type of desert; temperate deserts, also referred to as cold deserts, or hot or subtropical deserts. Hot deserts and cold deserts have different kinds of ecosystems. However, despite being very different, the two kinds of deserts have a few similarities.

Similarities of temperate and subtropical deserts:

- Both get fewer than 10 inches of rain annually.

- Dry air is found in both kinds of biomes.

- Both have harsh living conditions that impact people or animals living there.

- Plants have adapted to having less water and harsher temperatures.

- Animals have adapted to the conditions as far as energy, food consumption and when to get out and be active.

In general, deserts are made up of a number of abiotic components – including sand, the lack of moisture, and hot temperatures – basically anything that makes up an eco-system that isn't alive. However, there are also a number of biotic factors that affect deserts, which include living things, such as plants and animals.

Abiotic Components

Temperate Deserts

Antarctica is an example of a temperate desert. The temperatures are actually so cold, they could lead to the death of humans. In order to survive, the animals that live in these kinds of deserts have adapted with the passage of time. The ways they have done this is by adding extra layers of fat, or needing less food and energy in order to survive.

Subtropical Deserts

These deserts are too hot for many plants and animals to handle. The animals who call these deserts home have adapted to having less water. Because it is so hot during the day, they have become nocturnal, getting out during the night when it is cooler and easier to maneuver without getting overheated. But, because the nights are cold, they have had to become accustomed to the colder nights. Plants have had to adjust to having less water, so they are sparse and often close to the ground.

Location

Mountains: There are two major factors in the deserts' creation; mountains' rain shadows and the large circulation of global winds. As water-filled air is pushed up the mountain slopes, it cools then drops water on that particular side of the mountain. In the event of larger mountain ranges, very little water makes it to the other side. Therefore deserts are often found near mountainous areas, such as:

- The Caucasus Mountains in Asia, where the Karakum and Kyzyl Kum deserts are;

- The Atacama Desert, which is partly caused by the Andes Mountains in Chile;

- Parts of California, where the Santa Cruz mountains are;

- The Sahara desert, which is affected by a number of different mountain ranges.

Wind patterns: Global wind patterns, which are complicated, play a significant role in where deserts are located. Winds that circle the globe are the result of the difference between warmer equatorial temperatures as well as the polar temperatures that are cooler. After air has been warmed at the equator, it moves upward. Then it moves toward the north pole and toward the south pole, where it loses moisture, cools off and then sinks before returning to the equator. Therefore, stable wind patterns and shifting global patterns can contribute to where a desert is.

The passage of time greatly influences where and how deserts form. As time has passed, the locations of deserts have moved through the passage of geologic time. This change has been the result of the uplifting of mountain ranges and the continental drift. The horse latitudes are where more deserts are situated, which is generally straddling the Tropic of Capricorn and the Tropic of Cancer, which falls between 15 and 30 degrees to the equator's north.

There are geologically ancient deserts, such as the Sahara Desert in northern Africa, which is 65 million years old or the Kalahari in central Africa. In North America, three of the four major deserts are within a geological region called the Range Province and the Basin, which falls between the Sierra Nevadas and the Rocky Mountains then extending into the state of Sonora in Mexico.

The forces of erosion thousands of years past shaped the desert landscapes during heavy rainfall. The rocky mountain slopes and hillsides caught the rain, which picked up loose sediment, sand, cobbles and boulders then moved them. As gravity caused the water to be carried downhill, sediment was moved down to the basin. At the bottom of the mountain, the water spread out across a broad area where the mouths of canyons were widened.

Temperature

The temperature of a given desert will vary due to its geographic location. However, a characteristic of all deserts is the dryness. Heat is reflected by water vapor, which is either in the form of cloud cover or humidity, resulting in a cooling effect. Because of the reactions and the characteristics, deserts experience extreme temperatures, regardless of whether it is heat or cold.

The temperature fluctuations can result in other effects. Cool air sinks and warm air rises, so the fast changes of temperatures cause the air to move fast from one place to another. Because of that, deserts are windy, and those conditions contribute to evaporation. About 90% of available sunlight is transmitted by clear dry air, which in comparison to a typical humid climate seeing 40% of the available sunlight. The additional sunlight has ultraviolet radiation, which can cause major damage to plants, animals and people.

Precipitation

The desert environment has an unpredictable and uneven of the precipitation that is does receive, although that precipitation is minimal in nature. Precipitation amounts can vary from year to year. Some years it may seem as though the desert has gotten more rainfall than usual, but most years have very little rainfall. There can actually be entire years that the desert doesn't see a drop of rain.

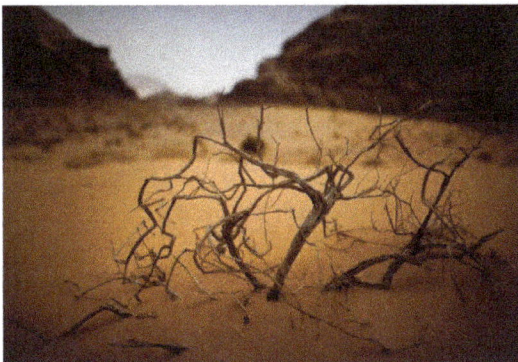

Biotic Components

Plant Life

Water is important everywhere and for every living thing. And it is, of course, extremely important in the desert. Because of the lack of water, the plants have made major adaptations.

Plant Adaptations

- The seeds of annual plants stay dormant until a time when there is adequate rainfall available to support a young plant.

- Cacti and other succulent plants store water in their spines, which are residual leaves. The stem is where photosynthesis takes place and the stem has pleats that are able to expand fast when rain falls.

- Evergreens have way cuticles and sunken stomata on shrubs that help hold water and prevent it from escaping. As an example, the holly plant's leaves are held at 70-degree angles so the sun only hits its sides. When the sun sinks low in the sky, the entire leaf is exposed. A fine salt covering is on the leaves and that helps reflect the sun off of the plant.

More than a fifth of the earth's land is comprised of deserts. The lack of water can create a survival problem for any humans, animals, plants or organisms. Besides the low rainfall, deserts experience a high amount of water loss from evaporation from the ground and through transpiration of plants. Evapotranspiration is from the combination of evaporation and transpiration. Potential evapotranspiration is how much water that would be lost by transpiration and evaporation if they were possible. Scientists measure this amount under controlled conditions with a large pan of water.

Soil in the desert is known for its coarseness, which permits the little moisture that is in it to pass through quickly, which means it is not as available for plants. Salts accumulate as a result from the high evaporation rate. The soil becomes alkaline and limits plant growth, which is also known as primary productivity.

Animal Life

Because of the entire process required to maintain life in the desert, the impact is that the size of individual animals is limited as well as the size of animal populations. The extremes of heat and aridity result in deserts being one of the most fragile of the ecosystems in the world.

Visitors to the desert should also take the proper precautions to protect themselves as the environment is much different than any other location.

Despite common beliefs that things can't live in the desert there are a number of creatures that have learned to survive on the distinctive plant life and in the difficult conditions.

- Large mammals like camels make their homes in the desert, and are suited to travel long periods of time without water. Lions live in the deserts of Africa, although they are endangered due to changing weather patterns and the presence of humans.

- Small rodents find homes in the desert, with variations from gerbils to hedgehogs. Larger hyenas and jackals are also often found in deserts.

- Lizards and snakes are particularly suited to the dry, hot climate of the desert, as are amphibious creatures like a number of toads and salamanders.

Aquatic Ecosystem

An aquatic ecosystem includes a group of interacting organisms which are dependent on one another and their water environment for nutrients and shelter. Examples of aquatic ecosystem include oceans, lakes and rivers.

An aquatic ecosystem includes freshwater habitats like lakes, ponds, rivers, oceans and streams, wetlands, swamp, etc. and marine habitats include oceans, intertidal zone, reefs, seabed and so on. The aquatic ecosystem is the habitat for water-dependent living species including animals, plants, and microbes.

Types of Aquatic Ecosystem

Different types of aquatic ecosystems are as follows:

Freshwater Aquatic Ecosystem

They cover only a small portion of earth nearly 0.8 per cent. Freshwater involves lakes, ponds, rivers and streams, wetlands, swamp, bog and temporary pools. Freshwater habitats are classified into lotic and lentic habitats. Water bodies such as lakes, ponds, pools, bogs, and other reservoirs are standing water and known as lentic habitats. Whereas lotic habitats represent flowing water bodies such as rivers, streams.

- Lotic Ecosystems: They mainly refer to the rapidly flowing waters that move in a unidirectional way including the rivers and streams. These environments harbor numerous species of insects such as beetles, mayflies, stoneflies and several species of fishes including trout, eel, minnow, etc. Apart from these aquatic species, these ecosystems also include various mammals such as beavers, river dolphins and otters.

- Lentic Ecosystems: They include all standing water habitats. Lakes and ponds are the main examples of Lentic Ecosystem. The word lentic mainly refers to stationary or relatively still water. These ecosystems are home to algae, crabs, shrimps, amphibians such as frogs and salamanders, for both rooted and floating-leaved plants and reptiles including alligators and other water snakes are also found here.

- Wetlands: Wetlands are marshy areas and are sometimes covered in water which has a wide diversity of plants and animals. Swamps, marshes, bogs, black spruce and water lilies are some examples in the plant species found in the wetlands. The animal life of this ecosystem consists of dragonflies and damselflies, birds such as Green Heron and fishes such as Northern Pike.

Marine Aquatic Ecosystem

Marine ecosystem covers the largest surface area of the earth. Two third of earth is covered by water and they constitute of oceans, seas, intertidal zone, reefs, seabed, estuaries, hydrothermal vents and rock pools. Each life form is unique and native to its habitat. This is because they have adaptations according to their habitat. In the case of aquatic animals, they can't survive outside of water. Exceptional cases are still there which shows another example of adaptations (e.g. mudskippers). The marine ecosystem is more concentrated with salts which make it difficult for freshwater organisms to live in. Also, marine animals cannot survive in freshwater. Their body is adapted to live in saltwater; if they are placed in less salty water, their body will swell (osmosis).

- Ocean Ecosystems: Our planet earth is gifted with the five major oceans, namely Pacific, Indian, Arctic, and the Atlantic Ocean. Among all these five oceans, the Pacific and the Atlantic are the largest and deepest ocean. These oceans serve as a home to more than five lakh aquatic species. Few creatures of these ecosystems include shellfish, shark, tube worms, crab small and large ocean fishes, turtles, crustaceans, blue whale, reptiles, marine mammals, seabirds, plankton, corals and other ocean plants.

- Coastal Systems: They are the open systems of land and water which are joined together to form the coastal ecosystems. The coastal ecosystems have a different structure, and diversity. A wide variety of species of aquatic plants and algae are found at the bottom of the coastal ecosystem. The fauna is diverse and it mainly consists of crabs, fish, insects, lobsters snails, shrimp, etc.

Plants and animals in an aquatic ecosystem show a wide variety of adaptations which may involve life cycle, physiological, structural and behavioral adaptations. Majority of aquatic animals are streamlined which helps them to reduce friction and thus save energy. Fins and gills are the locomotors and respiratory organs respectively. Special features in freshwater organisms help them to drain excess water from the body. Aquatic plants have different types of roots which help them to survive in water. Some may have submerged roots; some have emergent roots or maybe floating plants like water hyacinths.

Grasslands

Grasslands are areas where the vegetation is dominated by grasses (Poaceae); however, sedge (Cyperaceae) and rush (Juncaceae) families can also be found along with variable proportions of legumes, like clover, and other herbs. Grasslands occur naturally on all continents except Antarctica. Grasslands are found in most ecoregions of the Earth. For example, there are five terrestrial ecoregion classifications (subdivisions) of the temperate grasslands, savannas, and shrublands biome (ecosystem), which is one of eight terrestrial ecozones.

Vegetation

Grassland vegetation can vary in height from very short, as in chalk grassland, to quite tall, as in the case of North American tallgrass prairie, South American grasslands and African savanna.

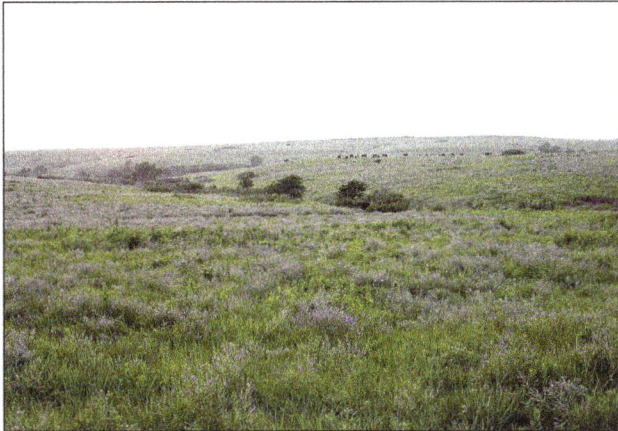

The Konza tallgrass prairie in the Flint Hills.

Woody plants, shrubs or trees may occur on some grasslands – forming savannas, scrubby grassland or semi-wooded grassland, such as the African savannas or the Iberian dehesa.

As flowering plants and trees, grasses grow in great concentrations in climates where annual rainfall ranges between 500 and 900 mm (20 and 35 in). The root systems of perennial grasses and forbs form complex mats that hold the soil in place.

Climates

Grasslands often occur in areas with annual precipitation is between 600 mm (24 in) and 1,500 mm (59 in) and average mean annual temperatures ranges from −5 and 20 °C. However, some grasslands occur in colder (−20 °C) and hotter (30 °C) climatic conditions. Grassland can exist in habitats that are frequently disturbed by grazing or fire, as such disturbance prevents the encroachment of woody species. Species richness is particularly high in grasslands of low soil fertility such as serpentine barrens and calcareous grasslands, where woody encroachment is prevented as low nutrient levels in the soil may inhibit the growth of forest and shrub species.

Biodiversity and Conservation

Grasslands dominated by unsown wild-plant communities ("unimproved grasslands") can be called either natural or "semi-natural" habitat. The majority of grasslands in temperate climates are "semi-natural". Although their plant communities are natural, their maintenance depends upon anthropogenic activities such as low-intensity farming, which maintains these grasslands through grazing and cutting regimes. These grasslands contain many species of wild plants, including grasses, sedges, rushes, and herbs; 25 or more species per square meter is not unusual. Chalk downlands in England can support over 40 species per square meter. In many parts of the world, few examples have escaped agricultural improvement (fertilizing, weed killing, plowing or re-seeding). For example, original North American prairie grasslands or lowland wildflower meadows in the UK are now rare and their associated wild flora equally threatened. Associated with the wild-plant diversity of the "unimproved" grasslands is usually a rich invertebrate fauna; there are also many species of birds that are grassland "specialists", such as the snipe and the great bustard. Agriculturally improved grasslands, which dominate modern intensive agricultural landscapes, are usually poor in wild plant species due to the original diversity of plants having been destroyed by cultivation, the original wild-plant communities having been replaced by sown monocultures of cultivated varieties of grasses and clovers, such as perennial ryegrass and white clover. In many parts of the world, "unimproved" grasslands are one of the most threatened types of habitat, and a target for acquisition by wildlife conservation groups or for special grants to landowners who are encouraged to manage them appropriately.

Human Impact and Economic Importance

Grassland vegetation is often a plagioclimax; it remains dominant in a particular area usually due to grazing, cutting, or natural or man-made fires, all discouraging colonization by and survival of tree and shrub seedlings. Some of the world's largest expanses of grassland are found in the African savanna, and these are maintained by wild herbivores as well as by nomadic pastoralists and their cattle, sheep or goats. Recently, grasslands have been shown to have an impact on climate change by slower decomposition rates of litter compared to forest environments.

Grassland in Cantabria.

A restored grassland ecosystem.

Grasslands may occur naturally or as the result of human activity. Grasslands created and maintained by human activity are called anthropogenic grasslands. Hunting cultures around the world often set regular fires to maintain and extend grasslands, and prevent fire-intolerant trees and shrubs from taking hold. The tallgrass prairies in the U.S. Midwest may have been extended eastward into Illinois, Indiana, and Ohio by human agency. Much grassland in northwest Europe developed after the Neolithic Period when people gradually cleared the forest to create areas for raising their livestock.

The professional study of grasslands falls under the category of rangeland management, which focuses on ecosystem services associated with the grass-dominated arid and semi-arid rangelands of the world. Rangelands account for an estimated 70% of the earth's landmass; thus, many cultures including those of the United States are indebted to the economics that the world's grasslands have to offer, from producing grazing animals, tourism, ecosystems services such as clean water and air, and energy extraction.

Types of Grassland

Schimper

Grassland types by Schimper:

- Meadow (hygrophilous or tropophilous grassland);

- Steppe (xerophilous grassland);

- Savannah (xerophilous grassland containing isolated trees).

Ellenberg and Mueller-Dombois

Grassland types by Ellenberg and Mueller-Dombois:

Formation-class V. Terrestrial herbaceous communities.

- Savannas and related grasslands (tropical or subtropical grasslands and parklands);

- Steppes and related grasslands (e.g. North American "prairies" etc);

- Meadows, pastures or related grasslands;

- Sedge swamps and flushes;

- Herbaceous and half-woody salt swamps;

- Forb vegetation.

Laycock

Grassland types by Laycock:

- Tallgrass (true) prairie;

- Shortgrass prairie;

- Mixed-grass prairie;

- Shrub steppe;

- Annual grassland;

- Desert (arid) grassland;

- High mountain grassland.

Other

Tropical and Subtropical

These grasslands are classified with tropical and subtropical savannas and shrublands as the tropical and subtropical grasslands, savannas, and shrublands biome. Notable tropical and subtropical grasslands include the Llanos grasslands of South America.

Temperate

Mid-latitude grasslands, including the prairie and Pacific grasslands of North America, the Pampas of Argentina, Brazil and Uruguay, calcareous downland, and the steppes of Europe. They are classified with temperate savannas and shrublands as the temperate grasslands, savannas, and shrublands biome. Temperate grasslands are the home to many large herbivores, such as bison, gazelles, zebras, rhinoceroses, and wild horses. Carnivores like lions, wolves and cheetahs and leopards are also found in temperate grasslands. Other animals of this region include: deer, prairie dogs, mice, jack rabbits, skunks, coyotes, snakes, fox, owls, badgers, blackbirds (both Old and New World varieties), grasshoppers, meadowlarks, sparrows, quails, hawks and hyenas.

Negri-nepote temperate grasslands.

Flooded

Grasslands that are flooded seasonally or year-round, like the Everglades of Florida, the Pantanal of Brazil, Bolivia and Paraguay or the Esteros del Ibera in Argentina, are classified with flooded savannas as the flooded grasslands and savannas biome and occur mostly in the tropics and subtropics.

Watermeadows are grasslands that are deliberately flooded for short periods.

Montane

Grassland in the Antelope Valley.

High-altitude grasslands located on high mountain ranges around the world, like the Páramo of the Andes Mountains. They are part of the montane grasslands and shrublands biome and also constitute tundra.

Tundra Grasslands

Similar to montane grasslands, polar Arctic tundra can have grasses, but high soil moisture means that few tundras are grass-dominated today. However, during the Pleistocene glacial periods (commonly referred to as ice ages), a freezing grassland known as steppe-tundra or mammoth steppe occupied large areas of the Northern Hemisphere. These areas were very cold and arid and featured sub-surface permafrost (hence tundra) but were nevertheless productive grassland ecosystems supporting a wide variety of fauna. As the temperature warmed and the climate became wetter at the beginning of the Holocene much of the mammoth steppe transitioned forest, while the drier parts in central Eurasia remained grassland, becoming the modern Eurasian steppe.

Desert and Xeric

Also called desert grasslands, this is composed of sparse grassland ecoregions located in the deserts and xeric shrublands biome.

Animals

Mites, insect larvae, nematodes and earthworms inhabit deep soil, which can reach 6 metres (20 ft) underground in undisturbed grasslands on the richest soils of the world. These invertebrates, along with symbiotic fungi, extend the root systems, break apart hard soil, enrich it with urea and other natural fertilizers, trap minerals and water and promote growth. Some types of fungi make the plants more resistant to insect and microbial attacks.

Grassland in all its form supports a vast variety of mammals, reptiles, birds, and insects. Typical large mammals include the blue wildebeest, American bison, giant anteater and Przewalski's horse.

While grasslands in general support diverse wildlife, given the lack of hiding places for predators, the African savanna regions support a much greater diversity in wildlife than do temperate grasslands.

There is evidence for grassland being much the product of animal behaviour and movement; some examples include migratory herds of antelope trampling vegetation and African bush elephants eating acacia saplings before the plant has a chance to grow into a mature tree.

Taiga

Taiga is generally referred to in North America as boreal forest or snow forest, is a biome characterized by coniferous forests consisting mostly of pines, spruces, and larches.

The taiga or boreal forest is the world's largest land biome. In North America, it covers most of inland Canada, Alaska, and parts of the northern contiguous United States.

In Eurasia, it covers most of Sweden, Finland, much of Norway and Estonia, some of the Scottish Highlands, some lowland/coastal areas of Iceland, much of Russia from Karelia in the west to the Pacific Ocean (including much of Siberia), and areas of northern Kazakhstan, northern Mongolia, and northern Japan (on the island of Hokkaidō). However, the main tree species, the length of the growing season and summer temperatures vary. For example, the taiga of North America mostly consists of spruces; Scandinavian and Finnish taiga consists of a mix of spruce, pines and birch; Russian taiga has spruces, pines and larches depending on the region, while the Eastern Siberian taiga is a vast larch forest.

A different use of the term taiga is often encountered in the English language, with "boreal forest" used in the United States and Canada to refer to only the more southerly part of the biome, while "taiga" is used to describe the more barren areas of the northernmost part of the biome approaching the tree line and the tundra biome. Hoffman (1958) discusses the origin of this differential use in North America and why it is an inappropriate differentiation of the Russian term. Although at high elevations taiga grades into alpine tundra through Krummholz, it is not exclusively an alpine biome; and unlike subalpine forest, much of taiga is lowlands.

White spruce taiga in the Alaska Range.

Climate and Geography

Taiga is the world's largest land biome (depending on how one defines a biome, it could also be considered the second-largest, after deserts and xeric shrublands), covering 17 million square kilometres (6.6 million square miles) or 11.5% of the Earth's land area. The largest areas are located in Russia and Canada. The taiga is the terrestrial biome with the lowest annual average temperatures after the tundra and permanent ice caps. Extreme winter minimums in the northern taiga are typically lower than those of the tundra. The lowest reliably recorded temperatures in the Northern Hemisphere were recorded in the taiga of northeastern Russia. The taiga or boreal forest has a subarctic climate with very large temperature range between seasons, but the long and cold winter is the dominant feature. This climate is classified as *Dfc, Dwc, Dsc, Dfd* and *Dwd* in the Köppen climate classification scheme, meaning that the short summer (24 h average 10 °C (50 °F) or more) lasts 1–3 months and always less than 4 months. In Siberian taiga the average temperature of the coldest month is between −6 °C (21 °F) and −50 °C (−58 °F). There are also some much smaller areas grading towards the oceanic *Cfc* climate with milder winters, whilst the extreme south and (in Eurasia) west of the taiga reaches into humid continental climates (*Dfb*, *Dwb*) with longer summers. The mean annual temperature generally varies from −5 to 5 °C (23 to 41 °F), but there are taiga areas in eastern Siberia and interior Alaska-Yukon where the mean annual reaches down to −10 °C (14 °F). According to some sources, the boreal forest grades into a temperate mixed forest when mean annual temperature reaches about 3 °C (37 °F). Discontinuous permafrost is found in areas with mean annual temperature below freezing (0 °C; 32 °F), whilst in the *Dfd* and *Dwd* climate zones continuous permafrost occurs and restricts growth to very shallow-rooted trees like Siberian larch. The winters, with average temperatures below freezing, last five to seven months. Temperatures vary from −54 to 30 °C (−65 to 86 °F) throughout the whole year. The summers, while short, are generally warm and humid. In much of the taiga, −20 °C (−4 °F) would be a typical winter day temperature and 18 °C (64 °F) an average summer day.

The taiga in the river valley at 67 °N, experiences the coldest winter temperatures in the northern hemisphere, but the extreme continentality of the climate gives an average daily high of 22 °C (72 °F).

Boreal forest near Shovel Point in Tettegouche State Park, along the northern shore of Lake Superior.

The growing season, when the vegetation in the taiga comes alive, is usually slightly longer than the climatic definition of summer as the plants of the boreal biome have a lower threshold to trigger growth. In Canada, Scandinavia and Finland, the growing season is often estimated by using the period of the year when the 24-hour average temperature is +5 °C (41 °F) or more. For the Taiga Plains in Canada, growing season varies from 80 to 150 days, and in the Taiga Shield from 100 to 140 days. Some sources claim 130 days growing season as typical for the taiga. Other sources mention that 50–100 frost-free days are characteristic. Data for locations in southwest Yukon gives 80–120 frost-free days. The closed canopy boreal forest in Kenozersky National Park near Plesetsk, Arkhangelsk Province, Russia, on average has 108 frost-free days. The longest growing season is found in the smaller areas with oceanic influences; in coastal areas of Scandinavia and Finland, the growing season of the closed boreal forest can be 145–180 days. The shortest growing season is found at the northern taiga–tundra ecotone, where the northern taiga forest no longer can grow and the tundra dominates the landscape when the growing season is down to 50–70 days, and the 24-hr average of the warmest month of the year usually is 10 °C (50 °F) or less. High latitudes mean that the sun does not rise far above the horizon, and less solar energy is received than further south. But the high latitude also ensures very long summer days, as the sun stays above the horizon nearly 20 hours each day, with only around 6 hours of daylight occurring in the dark winters, depending on latitude. The areas of the taiga inside the Arctic Circle have midnight sun in mid-summer and polar night in mid-winter.

Lakes and other water bodies are common in the taiga. The Helvetinjärvi National Park, is situated in the closed canopy taiga (mid-boreal to south-boreal) with mean annual temperature of 4 °C (39 °F).

The taiga experiences relatively low precipitation throughout the year (generally 200–750 mm (7.9–29.5 in) annually, 1,000 mm (39 in) in some areas), primarily as rain during the summer months, but also as fog and snow. This fog, especially predominant in low-lying areas during and after the thawing of frozen Arctic seas, means that sunshine is not abundant in the taiga even during the long summer days. As evaporation is consequently low for most of the year, precipitation exceeds evaporation, and is sufficient to sustain the dense vegetation growth including large trees. (In the steppe biome, often found south of taiga in the northern hemisphere, evapotranspiration exceeds precipitation, restricting vegetation to mostly grasses).

Snow may remain on the ground for as long as nine months in the northernmost extensions of the taiga ecozone.

In general, taiga grows to the south of the 10 °C (50 °F) July isotherm, but occasionally as far north as the 9 °C (48 °F) July isotherm. Rich in spruces, Scots pines in the western Siberian plain, the taiga is dominated by larch in Eastern Siberia, before returning to its original floristic richness on the Pacific shores. Two deciduous trees mingle throughout southern Siberia: birch and Populus tremula.

Late September in the fjords. This oceanic part of the forest can
see more than 1,000 mm (39 in) precipitation annually
and has warmer winters than the vast inland taiga.

The southern limit is more variable, depending on rainfall; taiga may be replaced by forest steppe south of the 15 °C (59 °F) July isotherm where rainfall is very low, but more typically extends south to the 18 °C (64 °F) July isotherm, and locally where rainfall is higher (notably in eastern Siberia and adjacent Outer Manchuria) south to the 20 °C (68 °F) July isotherm. In these warmer areas the taiga has higher species diversity, with more warmth-loving species such as Korean pine, Jezo spruce, and Manchurian fir, and merges gradually into mixed temperate forest or, more locally (on the Pacific Ocean coasts of North America and Asia), into coniferous temperate rainforests where oak and hornbeam appear and join the conifers, birch and Populus tremula.

Several of the world's longest rivers go through the taiga.

The area currently classified as taiga in Europe and North America (except Alaska) was recently glaciated. As the glaciers receded they left depressions in the topography that have since filled with water, creating lakes and bogs (especially muskeg soil) found throughout the taiga.

In Sweden the taiga is associated with the Norrland terrain.

Soils

Taiga soil tends to be young and poor in nutrients. It lacks the deep, organically enriched profile present in temperate deciduous forests. The thinness of the soil is due largely to the cold, which hinders the development of soil and the ease with which plants can use its nutrients. Fallen leaves and moss can remain on the forest floor for a long time in the cool, moist climate, which limits their organic contribution to the soil; acids from evergreen needles further leach the soil, creating spodosol, also known as podzol. Since the soil is acidic due to the falling pine needles, the forest floor has only lichens and some mosses growing on it. In clearings in the forest and in areas with more boreal deciduous trees, there are more herbs and berries growing. Diversity of soil organisms in the boreal forest is high, comparable to the tropical rainforest.

Flora

Boreal forest near Lake Baikal.

Since North America and Asia used to be connected by the Bering land bridge, a number of animal and plant species (more animals than plants) were able to colonize both continents and are distributed throughout the taiga biome. Others differ regionally, typically with each genus having several distinct species, each occupying different regions of the taiga. Taigas also have some small-leaved deciduous trees like birch, alder, willow, and poplar; mostly in areas escaping the most extreme winter cold. However, the Dahurian larch tolerates the coldest winters in the Northern Hemisphere in eastern Siberia. The very southernmost parts of the taiga may have trees such as oak, maple, elm and lime scattered among the conifers, and there is usually a gradual transition into a temperate mixed forest, such as the eastern forest-boreal transition of eastern Canada. In the interior of the continents with the driest climate, the boreal forests might grade into temperate grassland.

There are two major types of taiga. The southern part is the closed canopy forest, consisting of many closely spaced trees with mossy ground cover. In clearings in the forest, shrubs and wildflowers are common, such as the fireweed. The other type is the lichen woodland or sparse taiga, with trees that are farther-spaced and lichen ground cover; the latter is common in the northernmost taiga. In the northernmost taiga the forest cover is not only more sparse, but often stunted in growth form; moreover, ice pruned asymmetric black spruce (in North America) are often seen, with diminished foliage on the windward side. In Canada, Scandinavia and Finland, the boreal forest is usually divided into three subzones: The high boreal (north boreal) or taiga zone; the middle boreal (closed forest); and the southern boreal, a closed canopy boreal forest with some scattered temperate deciduous trees among the conifers, such as maple, elm and oak. This southern boreal forest experiences the longest and warmest growing season of the biome, and in some regions (including Scandinavia, Finland and western Russia) this subzone is commonly used for agricultural purposes. The boreal forest is home to many types of berries; some are confined to the southern and middle closed boreal forest (such as wild strawberry and partridgeberry); others grow in most areas of the taiga (such as cranberry and cloudberry), and some can grow in both the taiga and the low arctic (southern part of) tundra (such as bilberry, bunchberry and lingonberry).

Taiga spruce forest. Trees in this environment tend to grow
closer to the trunk and not "bush out" in
the normal manner of spruce trees.

The forests of the taiga are largely coniferous, dominated by larch, spruce, fir and pine. The woodland mix varies according to geography and climate so for example the Eastern Canadian forests ecoregion of the higher elevations of the Laurentian Mountains and the northern Appalachian Mountains in Canada is dominated by balsam fir *Abies balsamea*, while further north the Eastern Canadian Shield taiga of northern Quebec and Labrador is notably black spruce *Picea mariana* and tamarack larch *Larix laricina*.

Evergreen species in the taiga (spruce, fir, and pine) have a number of adaptations specifically for survival in harsh taiga winters, although larch, which is extremely cold-tolerant, is deciduous. Taiga trees tend to have shallow roots to take advantage of the thin soils, while many of them seasonally alter their biochemistry to make them more resistant to freezing, called "hardening". The narrow conical shape of northern conifers, and their downward-drooping limbs, also help them shed snow.

Because the sun is low in the horizon for most of the year, it is difficult for plants to generate energy from photosynthesis. Pine, spruce and fir do not lose their leaves seasonally and are able to photosynthesize with their older leaves in late winter and spring when light is good but temperatures are still too low for new growth to commence. The adaptation of evergreen needles limits the water lost due to transpiration and their dark green color increases their absorption of sunlight. Although precipitation is not a limiting factor, the ground freezes during the winter months and plant roots are unable to absorb water, so desiccation can be a severe problem in late winter for evergreens.

Moss (*Ptilium crista-castrensis*) cover on the floor of taiga.

Although the taiga is dominated by coniferous forests, some broadleaf trees also occur, notably birch, aspen, willow, and rowan. Many smaller herbaceous plants, such as ferns and occasionally ramps grow closer to the ground. Periodic stand-replacing wildfires (with return times of between 20–200 years) clear out the tree canopies, allowing sunlight to invigorate new growth on the forest floor. For some species, wildfires are a necessary part of the life cycle in the taiga; some, e.g. jack pine have cones which only open to release their seed after a fire, dispersing their seeds onto the newly cleared ground; certain species of fungi (such as morels) are also known to do this. Grasses grow wherever they can find a patch of sun, and mosses and lichens thrive on the damp ground and on the sides of tree trunks. In comparison with other biomes, however, the taiga has low biological diversity.

Jack pine cones and morels after fire in a boreal forest.

Coniferous trees are the dominant plants of the taiga biome. A very few species in four main genera are found: the evergreen spruce, fir and pine, and the deciduous larch. In North America, one or two species of fir and one or two species of spruce are dominant. Across Scandinavia and western Russia, the Scots pine is a common component of the taiga, while taiga of the Russian Far East and Mongolia is dominated by larch.

Fauna

Brown bear, Kamchatka peninsula. Brown bears are among the largest and most widespread taiga omnivores.

The boreal forest, or taiga, supports a relatively small range of animals due to the harshness of the climate. Canada's boreal forest includes 85 species of mammals, 130 species of fish, and an estimated 32,000 species of insects. Insects play a critical role as pollinators, decomposers, and as a part of the food web. Many nesting birds rely on them for food in the summer months. The cold winters and short summers make the taiga a challenging biome for reptiles and amphibians, which depend on environmental conditions to regulate their body temperatures, and there are only a few species in the boreal forest including red-sided garter snake, common European adder, blue-spotted salamander, northern two-lined salamander, Siberian salamander, wood frog, northern leopard frog, boreal chorus frog, American toad, and Canadian toad. Most hibernate underground in winter. Fish of the taiga must be able to withstand cold water conditions and be able to adapt to life under ice-covered water. Species in the taiga include Alaska blackfish, northern pike, walleye, longnose sucker, white sucker, various species of cisco, lake whitefish, round whitefish, pygmy whitefish, Arctic lamprey, various grayling species, brook trout (including sea-run brook trout in the Hudson Bay area), chum salmon, Siberian taimen, lenok and lake chub.

The taiga is home to a number of large herbivorous mammals, such as moose and reindeer/caribou. Some areas of the more southern closed boreal forest also have populations of other deer species such as the elk (wapiti) and roe deer. The largest animal in the taiga is the wood bison, found in northern Canada, Alaska and has been newly introduced into the Russian far-east. Small mammals of the Taiga biome include rodent species including beaver, squirrel, North American porcupine and vole, as well as a small number of lagomorph species such as snowshoe hare and mountain hare. These species have adapted to survive the harsh winters in their native ranges. Some larger mammals, such as bears, eat heartily during the summer in order to gain weight, and then go into hibernation during the winter. Other animals have adapted layers of fur or feathers to insulate them from the cold. Predatory mammals of the taiga must be adapted to travel long distances in search of scattered prey or be able to supplement their diet with vegetation or other forms of food (such as raccoons). Mammalian predators of the taiga include Canada lynx, Eurasian lynx, stoat, Siberian weasel, least weasel, sable, American marten, North American river otter, European otter, American mink, wolverine, Asian badger, fisher, gray wolf, coyote, red fox, brown bear, American black bear, Asiatic black bear, polar bear (only small areas at the taiga – tundra ecotone) and Siberian tiger.

More than 300 species of birds have their nesting grounds in the taiga. Siberian thrush, white-throated sparrow, and black-throated green warbler migrate to this habitat to take advantage of the long summer days and abundance of insects found around the numerous bogs and lakes. Of the 300 species of birds that summer in the taiga only 30 stay for the winter. These are either carrion-feeding or large raptors that can take live mammal prey, including golden eagle, rough-legged buzzard (also known as the rough-legged hawk), and raven, or else seed-eating birds, including several species of grouse and crossbills.

Fire

The Funny River Fire in Alaska burned 193,597 acres (78,346 ha), mostly Black spruce taiga.

Fire has been one of the most important factors shaping the composition and development of boreal forest stands; it is the dominant stand-renewing disturbance through much of the Canadian boreal forest. The fire history that characterizes an ecosystem is its *fire regime*, which has 3 elements: (1) fire type and intensity (e.g., crown fires, severe surface fires, and light surface fires), (2) size of typical fires of significance, and (3) frequency or return intervals for specific land units. The average time within a fire regime to burn an area equivalent to the total area of an ecosystem is its *fire rotation* or *fire cycle*. However, as Heinselman noted, each physiographic site tends to have its own return interval, so that some areas are skipped for long periods, while others might burn two-times or more often during a nominal fire rotation.

The dominant fire regime in the boreal forest is high-intensity crown fires or severe surface fires of very large size, often more than 10,000 ha, and sometimes more than 400,000 ha. Such fires kill entire stands. Fire rotations in the drier regions of western Canada and Alaska average 50–100 years, shorter than in the moister climates of eastern Canada, where they may average 200 years or more. Fire cycles also tend to be long near the tree line in the subarctic spruce-lichen woodlands. The longest cycles, possibly 300 years, probably occur in the western boreal in floodplain white spruce.

Amiro calculated the mean fire cycle for the period 1980 to 1999 in the Canadian boreal forest (including taiga) at 126 years. Increased fire activity has been predicted for western Canada, but parts of eastern Canada may experience less fire in future because of greater precipitation in a warmer climate.

The Shanta Creek Fire began in a taiga area that had not had a major fire in over 130 years, and so was allowed to burn unchecked until it began to threaten populated areas.

The mature boreal forest pattern in the south shows balsam fir dominant on well-drained sites in eastern Canada changing centrally and westward to a prominence of white spruce, with black spruce and tamarack forming the forests on peats, and with jack pine usually present on dry sites except in the extreme east, where it is absent. The effects of fires are inextricably woven into the patterns of vegetation on the landscape, which in the east favour black spruce, paper birch, and jack pine over balsam fir, and in the west give the advantage to aspen, jack pine, black spruce, and birch over white spruce. Many investigators have reported the ubiquity of charcoal under the forest floor and in the upper soil profile. Charcoal in soils provided Bryson with clues about the forest history of an area 280 km north of the then current tree line at Ennadai Lake, District Keewatin, Northwest Territories.

Two lines of evidence support the thesis that fire has always been an integral factor in the boreal forest: (1) direct, eye-witness accounts and forest-fire statistics, and (2) indirect, circumstantial evidence based on the effects of fire, as well as on persisting indicators. The patchwork mosaic of forest stands in the boreal forest, typically with abrupt, irregular boundaries circumscribing homogenous stands, is indirect but compelling testimony to the role of fire in shaping the forest.

The fact is that most boreal forest stands are less than 100 years old, and only in the rather few areas that have escaped burning are there stands of white spruce older than 250 years. The prevalence of fire-adaptive morphologic and reproductive characteristics of many boreal plant species is further evidence pointing to a long and intimate association with fire. Seven of the ten most common trees in the boreal forest—jack pine, lodgepole pine, aspen, balsam poplar (*Populus balsamifera*), paper birch, tamarack, black spruce—can be classed as pioneers in their adaptations for rapid invasion of open areas. White spruce shows some pioneering abilities, too, but is less able than black spruce and the pines to disperse seed at all seasons. Only balsam fir and alpine fir seem to be poorly adapted to reproduce after fire, as their cones disintegrate at maturity, leaving no seed in the crowns.

The oldest forests in the northwest boreal region, some older than 300 years, are of white spruce occurring as pure stands on moist floodplains. Here, the frequency of fire is much less than on adjacent uplands dominated by pine, black spruce and aspen. In contrast, in the Cordilleran region, fire is most frequent in the valley bottoms, decreasing upward, as shown by a mosaic of young pioneer pine and broadleaf stands below, and older spruce–fir on the slopes above. Without fire, the boreal forest would become more and more homogeneous, with the long-lived white spruce gradually replacing pine, aspen, balsam poplar, and birch, and perhaps even black spruce, except on the peatlands.

Threats

Human Activities

Large areas of Siberia's taiga have been harvested for lumber since the collapse of the Soviet Union. Previously, the forest was protected by the restrictions of the Soviet Forest Ministry, but with the collapse of the Union, the restrictions regarding trade with Western nations have vanished. Trees are easy to harvest and sell well, so loggers have begun harvesting Russian taiga evergreen trees for sale to nations previously forbidden by Soviet law.

Plesetsk Cosmodrome is situated in the taiga.

In Canada, eight percent of the taiga is protected from development, the provincial government allows forest management to occur on Crown land under rigorous constraints.

The main forestry practice in the boreal forest of Canada is clearcutting, which involves cutting down most of the trees in a given area, then replanting the forest as a monocrop (one species of tree) the following season.

Some of the products from logged boreal forests include toilet paper, copy paper, newsprint, and lumber. More than 90% of boreal forest products from Canada are exported for consumption and processing in the United States.

Some of the larger cities situated in this biome are Murmansk, Arkhangelsk, Yakutsk, Anchorage, Yellowknife, Tromsø, Luleå, and Oulu.

Most companies that harvest in Canadian forests are certified by an independent third party agency such as the Forest Stewardship Council (FSC), Sustainable Forests Initiative (SFI), or the Canadian Standards Association (CSA). While the certification process differs between these groups, all of them include forest stewardship, respect for aboriginal peoples, compliance with local, provincial or national environmental laws, forest worker safety, education and training, and other environmental, business, and social requirements. The prompt renewal of all harvest sites by planting or natural renewal is also required.

Climate Change

Seney national wildlife refuge.

During the last quarter of the twentieth century, the zone of latitude occupied by the boreal forest experienced some of the greatest temperature increases on Earth. Winter temperatures have increased more than summer temperatures. The number of days with extremely cold temperatures (e.g., −20 to −40 °C (−4 to −40 °F)) has decreased irregularly but systematically in nearly all the boreal region, allowing better survival for tree-damaging insects. In summer, the daily low temperature has increased more than the daily high temperature. In Fairbanks, Alaska, the length of the frost-free season has increased from 60–90 days in the early twentieth century to about 120 days a century later. Summer warming has been shown to increase water stress and reduce tree growth in dry areas of the southern boreal forest in central Alaska, western Canada and portions of far eastern Russia. Precipitation is relatively abundant in Scandinavia, Finland, northwest Russia and eastern Canada, where a longer growth season (i.e. the period when sap flow is not impeded by frozen water) accelerate tree growth. As a consequence of this warming trend, the warmer parts of the boreal forests are susceptible to replacement by grassland, parkland or temperate forest.

In Siberia, the taiga is converting from predominantly needle-shedding larch trees to evergreen conifers in response to a warming climate. This is likely to further accelerate warming, as the evergreen trees will absorb more of the sun's rays. Given the vast size of the area, such a change has the potential to affect areas well outside of the region. In much of the boreal forest in Alaska, the growth of white spruce trees are stunted by unusually warm summers, while trees on some of the coldest fringes of the forest are experiencing faster growth than previously.

Lack of moisture in the warmer summers are also stressing the birch trees of central Alaska.

Insects

Recent years have seen outbreaks of insect pests in forest-destroying plagues: the spruce-bark beetle (*Dendroctonus rufipennis*) in Yukon and Alaska; the mountain pine beetle in British Columbia; the aspen-leaf miner; the larch sawfly; the spruce budworm (*Choristoneura fumiferana*); the spruce coneworm.

Pollution

The effect of sulphur dioxide on woody boreal forest species was investigated by Addison, who exposed plants growing on native soils and tailings to 15.2 µmol/m³ (0.34 ppm) of SO_2 on CO_2 assimilation rate (NAR). The Canadian maximum acceptable limit for atmospheric SO_2 is 0.34 ppm. Fumigation with SO_2 significantly reduced NAR in all species and produced visible symptoms of injury in 2–20 days. The decrease in NAR of deciduous species (trembling aspen [*Populus tremuloides*], willow [*Salix*], green alder [*Alnus viridis*], and white birch [*Betula papyrifera*]) was significantly more rapid than of conifers (white spruce, black spruce [*Picea mariana*], and jack pine [*Pinus banksiana*]) or an evergreen angiosperm (Labrador tea) growing on a fertilized Brunisol. These metabolic and visible injury responses seemed to be related to the differences in S uptake owing in part to higher gas exchange rates for deciduous species than for conifers. Conifers growing in oil sands tailings responded to SO_2 with a significantly more rapid decrease in NAR compared with those growing in the Brunisol, perhaps because of predisposing toxic material in the tailings. However, sulphur uptake and visible symptom development did not differ between conifers growing on the 2 substrates.

Acidification of precipitation by anthropogenic, acid-forming emissions has been associated with damage to vegetation and reduced forest productivity, but 2-year-old white spruce that were subjected to simulated acid rain (at pH 4.6, 3.6, and 2.6) applied weekly for 7 weeks incurred no statistically significant (P 0.05) reduction in growth during the experiment compared with the background control (pH 5.6) . However, symptoms of injury were observed in all treatments, the number of plants and the number of needles affected increased with increasing rain acidity and with time. Scherbatskoy and Klein found no significant effect of chlorophyll concentration in white spruce at pH 4.3 and 2.8, but Abouguendia and Baschak found a significant reduction in white spruce at pH 2.6, while the foliar sulphur content significantly greater at pH 2.6 than any of the other treatments.

Protection

Peat bog. Bogs and peatland are widespread in the taiga. They are home to a unique flora, and store vast amounts of carbon.

Many nations are taking direct steps to protect the ecology of the taiga by prohibiting logging, mining, oil and gas production, and other forms of development. In February 2010 the Canadian government established protection for 13,000 square kilometres of boreal forest by creating a new 10,700-square-kilometre park reserve in the Mealy Mountains area of eastern Canada and a 3,000-square-kilometre waterway provincial park that follows alongside the Eagle River from headwaters to sea.

Two Canadian provincial governments, Ontario and Quebec, introduced measures in 2008 that would protect at least half of their northern boreal forest. Although both provinces admitted it will take years to plan, work with Aboriginal and local communities and ultimately map out precise boundaries of the areas off-limits to development, the measures are expected to create some of the largest protected areas networks in the world once completed. Both announcements came the following year after a letter signed by 1,500 scientists called on political leaders to protect at least half of the boreal forest.

The taiga stores enormous quantities of carbon, more than the world's temperate and tropical forests combined, much of it in wetlands and peatland. In fact, current estimates place boreal forests as storing twice as much carbon per unit area as tropical forests.

Natural Disturbance

One of the biggest areas of research and a topic still full of unsolved questions is the recurring disturbance of fire and the role it plays in propagating the lichen woodland. The phenomenon of wildfire by lightning strike is the primary determinant of understory vegetation and because of this, it is considered to be the predominant force behind community and ecosystem properties in the lichen woodland. The significance of fire is clearly evident when one considers that understory vegetation influences tree seedling germination in the short term and decomposition of biomass and nutrient availability in the long term. The recurrent cycle of large, damaging fire occurs approximately every 70 to 100 years. Understanding the dynamics of this ecosystem is entangled with discovering the successional paths that the vegetation exhibits after a fire. Trees, shrubs, and lichens all recover from fire-induced damage through vegetative reproduction as well as invasion by propagules. Seeds that have fallen and become buried provide little help in re-establishment of a species. The reappearance of lichens is reasoned to occur because of varying conditions and light/nutrient availability in each different microstate. Several different studies have been done that have led to the formation of the theory that post-fire development can be propagated by any of four pathways: self replacement, species-dominance relay, species replacement, or gap-phase self replacement. Self replacement is simply the re-establishment of the pre-fire dominant species. Species-dominance relay is a sequential attempt of tree species to establish dominance in the canopy. Species replacement is when fires occur in sufficient frequency to interrupt species dominance relay. Gap-Phase Self-Replacement is the least common and so far has only been documented in Western Canada. It is a self replacement of the surviving species into the canopy gaps after a fire kills another species. The particular pathway taken after a fire disturbance depends on how the landscape is able to support trees as well as fire frequency. Fire frequency has a large role in shaping the original inception of the lower forest line of the lichen woodland taiga.

It has been hypothesized by Serge Payette that the spruce-moss forest ecosystem was changed into the lichen woodland biome due to the initiation of two compounded strong disturbances: large fire and the appearance and attack of the spruce budworm. The spruce budworm is a deadly insect to the spruce populations in the southern regions of the taiga. J.P. Jasinski confirmed this theory five years later stating "Their [lichen woodlands] persistence, along with their previous moss forest histories and current occurrence adjacent to closed moss forests, indicate that they are an alternative stable state to the spruce–moss forests".

Tundra

In physical geography, tundra is a type of biome where the tree growth is hindered by low temperatures and short growing seasons. Tundra vegetation is composed of dwarf shrubs, sedges and grasses, mosses, and lichens. Scattered trees grow in some tundra regions. The ecotone (or ecological boundary region) between the tundra and the forest is known as the tree line or timberline.

There are three regions and associated types of tundra: Arctic tundra, alpine tundra, and Antarctic tundra.

Arctic

Arctic tundra occurs in the far Northern Hemisphere, north of the taiga belt. The word "tundra" usually refers only to the areas where the subsoil is permafrost, or permanently frozen soil. (It may also refer to the treeless plain in general, so that northern Sápmi would be included). Permafrost tundra includes vast areas of northern Russia and Canada. The polar tundra is home to several peoples who are mostly nomadic reindeer herders, such as the Nganasan and Nenets in the permafrost area (and the Sami in Sápmi).

Tundra in Siberia.

Arctic tundra contains areas of stark landscape and is frozen for much of the year. The soil there is frozen from 25 to 90 cm (10 to 35 in) down, making it impossible for trees to grow. Instead, bare and sometimes rocky land can only support certain kinds of

Arctic vegetation, low growing plants such as moss, heath (Ericaceae varieties such as crowberry and black bearberry), and lichen.

There are two main seasons, winter and summer, in the polar tundra areas. During the winter it is very cold and dark, with the average temperature around −28 °C (−18 °F), sometimes dipping as low as −50 °C (−58 °F). However, extreme cold temperatures on the tundra do not drop as low as those experienced in taiga areas further south (for example, Russia's and Canada's lowest temperatures were recorded in locations south of the tree line). During the summer, temperatures rise somewhat, and the top layer of seasonally-frozen soil melts, leaving the ground very soggy. The tundra is covered in marshes, lakes, bogs and streams during the warm months. Generally daytime temperatures during the summer rise to about 12 °C (54 °F) but can often drop to 3 °C (37 °F) or even below freezing. Arctic tundras are sometimes the subject of habitat conservation programs. In Canada and Russia, many of these areas are protected through a national Biodiversity Action Plan.

Vuntut National Park.

Tundra tends to be windy, with winds often blowing upwards of 50−100 km/h (30−60 mph). However, in terms of precipitation, it is desert-like, with only about 150−250 mm (6−10 in) falling per year (the summer is typically the season of maximum precipitation). Although precipitation is light, evaporation is also relatively minimal. During the summer, the permafrost thaws just enough to let plants grow and reproduce, but because the ground below this is frozen, the water cannot sink any lower, and so the water forms the lakes and marshes found during the summer months. There is a natural pattern of accumulation of fuel and wildfire which varies depending on the nature of vegetation and terrain. Research in Alaska has shown fire-event return intervals (FRIs) that typically vary from 150 to 200 years, with dryer lowland areas burning more frequently than wetter highland areas.

The biodiversity of tundra is low: 1,700 species of vascular plants and only 48 species of land mammals can be found, although millions of birds migrate there each year for the marshes. There are also a few fish species. There are few species with large populations. Notable animals in the Arctic tundra include reindeer (caribou), musk ox, Arctic hare,

Arctic fox, snowy owl, lemmings, and even polar bears near the ocean. Tundra is largely devoid of poikilotherms such as frogs or lizards.

A group of muskoxen in Alaska.

Due to the harsh climate of Arctic tundra, regions of this kind have seen little human activity, even though they are sometimes rich in natural resources such as petroleum, natural gas and uranium. In recent times this has begun to change in Alaska, Russia, and some other parts of the world: for example, the Yamalo-Nenets Autonomous Okrug produces 90% of Russia's natural gas.

Relationship to Global Warming

A severe threat to tundra is global warming, which causes permafrost to melt. The melting of the permafrost in a given area on human time scales (decades or centuries) could radically change which species can survive there.

Another concern is that about one third of the world's soil-bound carbon is in taiga and tundra areas. When the permafrost melts, it releases carbon in the form of carbon dioxide and methane, both of which are greenhouse gases. The effect has been observed in Alaska. In the 1970s the tundra was a carbon sink, but today, it is a carbon source. Methane is produced when vegetation decays in lakes and wetlands.

The amount of greenhouse gases which will be released under projected scenarios for global warming have not been reliably quantified by scientific studies.

In locations where dead vegetation and peat has accumulated, there is a risk of wildfire, such as the 1,039 km² (401 sq mi) of tundra which burned in 2007 on the north slope of the Brooks Range in Alaska. Such events may both result from and contribute to global warming.

Antarctic

Antarctic tundra occurs on Antarctica and on several Antarctic and subantarctic islands, including South Georgia and the South Sandwich Islands and the Kerguelen

Islands. Most of Antarctica is too cold and dry to support vegetation, and most of the continent is covered by ice fields. However, some portions of the continent, particularly the Antarctic Peninsula, have areas of rocky soil that support plant life. The flora presently consists of around 300–400 lichens, 100 mosses, 25 liverworts, and around 700 terrestrial and aquatic algae species, which live on the areas of exposed rock and soil around the shore of the continent. Antarctica's two flowering plant species, the Antarctic hair grass (*Deschampsia antarctica*) and Antarctic pearlwort (*Colobanthus quitensis*), are found on the northern and western parts of the Antarctic Peninsula. In contrast with the Arctic tundra, the Antarctic tundra lacks a large mammal fauna, mostly due to its physical isolation from the other continents. Sea mammals and sea birds, including seals and penguins, inhabit areas near the shore, and some small mammals, like rabbits and cats, have been introduced by humans to some of the subantarctic islands. The Antipodes Subantarctic Islands tundra ecoregion includes the Bounty Islands, Auckland Islands, Antipodes Islands, the Campbell Island group, and Macquarie Island. Species endemic to this ecoregion include *Nematoceras dienemum* and *Nematoceras sulcatum*, the only subantarctic orchids; the royal penguin; and the Antipodean albatross.

Tundra on the Kerguelen Islands.

There is some ambiguity on whether Magellanic moorland, on the west coast of Patagonia, should be considered tundra or not. Phytogeographer Edmundo Pisano called it tundra (Spanish: *Tundra Magallánica*) since he considered the low temperatures key to restrict plant growth.

The flora and fauna of Antarctica and the Antarctic Islands (south of 60° south latitude) are protected by the Antarctic Treaty.

Alpine

Alpine tundra does not contain trees because the climate and soils at high altitude block tree growth. The cold climate of the alpine tundra is caused by the low air temperatures, and is similar to polar climate. Alpine tundra is distinguished from arctic tundra in that alpine tundra typically does not have permafrost, and alpine soils are generally better drained than arctic soils. Alpine tundra transitions to subalpine forests below the tree line; stunted forests occurring at the forest-tundra ecotone (the treeline) are known as *Krummholz*.

Alpine tundra.

Alpine tundra occurs in mountains worldwide. The flora of the alpine tundra is characterized by plants that grow close to the ground, including perennial grasses, sedges, forbs, cushion plants, mosses, and lichens. The flora is adapted to the harsh conditions of the alpine environment, which include low temperatures, dryness, ultraviolet radiation, and a short growing season.

Climatic Classification

Tundra region with fjords, glaciers and mountains. Kongsfjorden, Spitsbergen.

Tundra climates ordinarily fit the Köppen climate classification *ET*, signifying a local climate in which at least one month has an average temperature high enough to melt snow (0 °C (32 °F)), but no month with an average temperature in excess of 10 °C (50 °F). The cold limit generally meets the *EF* climates of permanent ice and snows; the warm-summer limit generally corresponds with the poleward or altitudinal limit of trees, where they grade into the subarctic climates designated *Dfd*, *Dwd* and *Dsd* (extreme winters as in parts of Siberia), *Dfc* typical in Alaska, Canada, parts of Scandinavia, European Russia, and Western Siberia (cold winters with months of freezing), or even *Cfc* (no month colder than −3 °C (27 °F) as in parts of Iceland and southernmost South America). Tundra climates as a rule are hostile to woody vegetation even where the winters are comparatively mild by polar standards, as in Iceland.

Nenets people are nomadic reindeer herders.

Despite the potential diversity of climates in the *ET* category involving precipitation, extreme temperatures, and relative wet and dry seasons, this category is rarely subdivided. Rainfall and snowfall are generally slight due to the low vapor pressure of water in the chilly atmosphere, but as a rule potential evapotranspiration is extremely low, allowing soggy terrain of swamps and bogs even in places that get precipitation typical of deserts of lower and middle latitudes. The amount of native tundra biomass depends more on the local temperature than the amount of precipitation.

Montane Ecosystems

The Montane ecosystem has the richest diversity of plant and animal life. Meandering rivers and open meadows are surrounded by hilly slopes. Wildflowers blanket the meadows throughout the summer growing season.

Dry, south-facing slopes of the Montane often have open stands of large ponderosa pines. Spacing of ponderosa pines is somewhat related to available soil moisture. Grasses, other herbs and shrubs may grow between the widely spaced trees on dry slopes. As the pines become old, their bark changes from gray-brown to cinnamon-red, and the bark releases a pleasant fragrance when warmed by the sun. The long needles of ponderosa pines are attached to the stems in groups of two's and three's.

Ponderosa pine bark turns red as the tree ages.

North-facing slopes of the Montane escape some of the sun's drying action, so their soils contain more available water. As a result, the trees grow closer together and competition for sunlight produces a tall, slender growth form. The trees may be a mixture of Douglas fir, lodgepole pine, ponderosa pine and an occasional Engelmann spruce. A few shade-tolerant plants grow on the floor of the forest.

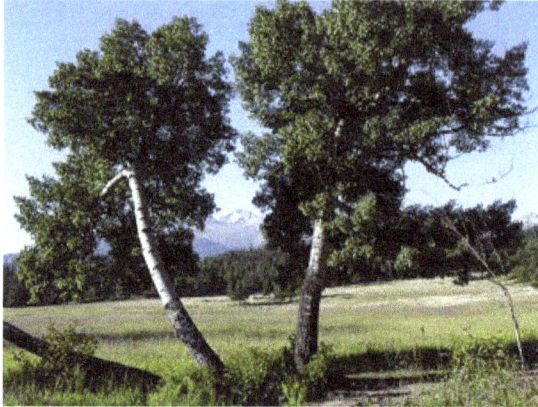

Old aspen trees in upper beaver meadows

Montane soils with high moisture content may support groves of quaking aspen, whose leaves turn golden-yellow in the autumn and whose whitish bark is easy to recognize. Along streams or the shores of lakes, other water-loving small trees may be found. These include various willows, mountain alder, and water birch with dark-colored bark. In a few places, blue spruce may grow near streams and sometimes hybridize with Engelmann spruce. Flat Montane valleys may frequently have water-logged soil and be unable to support growth of evergreen forests.

References

- Ecosystem, science: britannica.com, Retrieved 21 May, 2019

- Forest, entry: newworldencyclopedia.org, Retrieved 8 January, 2019

- "Forest regions temperate zone". Yale school of forestry and environmental studies. Retrieved 2019-07-17

- Aragao, l. E. O. C. (2009). "above- and below-ground net primary productivity across ten amazonian forests on contrasting soils". Biogeosciences. 6 (12): 2759–2778. Doi:10.5194/bg-6-2759-2009

- Why the amazon rainforest is so rich in species archived 25 february 2011 at the wayback machine. Earthobservatory.nasa.gov (5 december 2005). Retrieved on 28 march 2013

- Desert-ecosystem: conserve-energy-future.com, Retrieved 13 May, 2019

- Ruckstuhl, k. E.; johnson, e. A.; miyanishi, k. (july 2008). "boreal forest and global change". Philos. Trans. R. Soc. Lond. B biol. Sci. 363 (1501): 2245–9. Doi:10.1098/rstb.2007.2196. Pmc 2387060. Pmid 18006417

- Marsden, william (2008-11-16). "charest promises to protect north". Montreal gazette. Archived from the original on 5 april 2011. Retrieved 25 june 2012

Ecology of Plant   3

- **Plant Ecology**
- **Plant Community**
- **Plant Perception**
- **Plant Stress**
- **Plant Stress Measurement**

Plant ecology is concerned with the study of relationships between plants and their physical and biotic environment. Some of its focus areas include plant community, plant perception and plant stress measurement. The topics elaborated in this chapter will help in gaining a better perspective about plant ecology.

Plant Ecology

Plant ecology examines the relationships between plants and their physical and biotic environment. Plants are mostly sessile and photosynthetic organisms, and must attain their light, water, and nutrient resources directly from the immediate environment. Plant size and position in the community affect the capture and utilization of these resources and hence plants have evolved specific adaptations to enhance these capabilities. Understory plants have evolved mechanisms that allow them to tolerate low light conditions, while plants in the open have different mechanisms to cope with excess light. The absorption by roots and movement of water in the plant are determined by gradients in potential energy between the soil and atmosphere, as well as within the plant, as expressed by the concept of water potential. Nutrients are available through biological and chemical processes in the soil. Mycorrhizae are critical in absorption of phosphorus and are also capable of interconnecting plants through their hyphae, thus facilitating belowground transfers of nutrients and water. Plants possess various adaptive functions, such as different photosynthetic pathways, that provide greater fitness in certain environments. In addition, there are correlations among plant traits, such as a positive relationship between photosynthetic rate and leaf nitrogen, or between leaf mass per area and photosynthesis, which suggest that there are ecological rules governing functional traits that cross species lines. Resource competition occurs when one or more resources are in limited supply and plants have various adaptations that maximize competitive success, including. Allelopathy, when one plant releases an organic material into the environment to the detriment of a second plant. Plants also greatly influence the belowground environment (the rhizosphere) by altering the composition of the microbial community of bacteria and fungi. Interactions between above and belowground processes affect competitive outcomes and can alter community dynamics, including the process of successional change. Primary succession occurs on new substrate and secondary succession occurs where vegetation previously existed. Secondary successions are initiated by disturbances such as fire, wind damage, flooding, grazing, and disease. Disturbance frequency and intensity greatly determine the development of the plant community and current and future climate change may result in new communities not present under present conditions, nor that resemble any from the recent past, making predictions of such impacts difficult.

Plant Community

A plant community (sometimes "phytocoenosis" or "phytocenosis") is a collection or association of plant species within a designated geographical unit, which forms a relatively uniform patch, distinguishable from neighboring patches of different vegetation types. The components of each plant community are influenced by soil type, topography, climate and human disturbance. In many cases there are several soil types within a given phytocoenosis.

Alpine Heathland plant community at High Shelf Camp.

A plant community can be described floristically (the species it contains) and physiognomically (its physical structure). For example, a forest (a community of trees) includes the overstory, or upper tree layer of the canopy, as well as the understory, further subdivided into the shrub layer, herbaceous layer, and sometimes also moss layer. In some cases of complex forests there is also a well-defined lower tree layer. A plant community is similar in concept to a vegetation type, with the former having more of an emphasis on the ecological association of species within it, and the latter on overall appearance by which it is readily recognized by a layperson.

A plant community can be rare even if none of the major species defining it are rare. This is because it is the association of species and relationship to their environment that may be rare. An example is the Sycamore Alluvial Woodland in California dominated by the California sycamore Platanus racemosa. The community is rare, being localized to a small area of California and existing nowhere else, yet the California sycamore is not a rare tree in California.

An example is a grassland on the northern Caucasus Steppes, where common grass species found are Festuca sulcata and Poa bulbosa. A common sedge in this grassland phytocoenosis is Carex shreberi. Other representative forbs occurring in these steppe grasslands are Artemisia austriaca and Polygonum aviculare.

An example of a three tiered plant community is in Central Westland of South Island, New Zealand. These forests are the most extensive continuous reaches of podocarp/broadleaf forests in that country. The overstory includes miro, rimu and mountain totara. The mid-story includes tree ferns such as Cyathea smithii and Dicksonia squarrosa, whilst the lowest tier and epiphytic associates include Asplenium polyodon, Tmesipteris tannensis, Astelia solandri and Blechnum discolor.

Plant Perception

Plant perception or plant gnosophysiology is the ability of plants to sense and respond to the environment by adjusting their morphology, physiology, and phenotype accordingly. Botanical research has revealed that plants are capable of reacting to a broad variety of stimuli, including chemicals, gravity, light, moisture, infections, temperature, oxygen and carbon dioxide concentrations, parasite infestation, disease, physical disruption, sound, and touch. The scientific study of plant perception is informed by numerous disciplines, such as plant physiology, ecology, and molecular biology.

Processes

Detection

Positional displacement can be detected by plants. Poplar stems can detect reorientation and inclination (equilibrioception).

Pathway Signals

Wounded tomatoes are known to produce the volatile odour methyl jasmonate as an alarm signal. Neighbouring plants can then detect the chemical and prepare for attack by producing chemicals which defend against insects or attract insect predators. Plants systematically use hormonal signalling pathways to coordinate their own development and morphology.

Neurochemicals

Plants produce several proteins also found in animal neuron systems, such as acetylcholine esterase, glutamate receptors, GABA receptors, and endocannabinoid signaling components. They may also use ATP, NO, and ROS for signaling in the same ways that animals do.

Electrophysiology

Plants have a variety of methods of delivering electrical signals. The four commonly recognized propagation methods include action potentials (APs), variation potentials (VPs), local electric potentials (LEPs), and systemic potentials (SPs).

Although plant cells are not neurons, they can be electrically excitable and can display rapid electrical responses in the form of APs to environmental stimuli. APs allow for the movement of signaling ions and molecules from the pre-potential cell to the post-potential cell(s). These electrophysiological signals are constituted by gradient fluxes of ions such as H^+, K^+, Cl^-, Na^+, and Ca^{2+} but it is also thought that other electrically charge ions such as Fe^{3+}, Al^{3+}, Mg^{2+}, Zn^{2+}, Mn^{2+}, and Hg^{2+} may also play a role in downstream outputs. The maintenance of each ions electrochemical gradient is vital in the health of the cell in that if the cell would ever reach equilibrium with its environment, it is dead. This dead state can be due to a variety of reasons such as ion channel blocking or membrane puncturing.

These electrophysiological ions bind to receptors on the receiving cell causing downstream effects result from one or a combination of molecules present. This means of transferring information and activating physiological responses via a signaling molecule system has been found to be faster and more frequent in the presence of APs.

These action potentials can influence processes such as actin-based cytoplasmic streaming, plant organ movements, wound responses, respiration, photosynthesis, and flowering. These electrical responses can cause the synthesis of numerous organic molecules, including ones that act as neuroactive substances in other organisms.

The ion flux across cells also influence the movement of other molecules and solutes. This changes the osmotic gradient of the cell, resulting in changes to turgor pressure in plant cells by water and solute flux across cell membranes. These variations are vital for nutrient uptake, growth, many types of movements among other basic plant physiology and behavior.

Thus, plants achieve behavioural responses in environmental, communicative, and ecological contexts.

Signal Response

A plant's concomitant reactive behavior is mediated by phytochromes, kinins, hormones, antibiotic or other chemical release, changes of water and chemical transport, and other means. These responses are generally slow, taking at minimum a number of hours to accomplish, and can best be observed with time-lapse cinematography, but rapid movements can occur as well. Plants respond to volatile signals produced by other plants. Jasmonate levels also increase rapidly in response to mechanical perturbations such as tendril coiling.

Plants have many strategies to fight off pests. For example, they can produce a slew of different chemical toxins (phytoalexins) against predators and parasites or they can induce rapid cell death in invading cells to hinder the pests from spreading out.

Some plants are capable of rapid movement: the so-called "sensitive plant" (*Mimosa pudica*) responds to even the slightest physical touch by quickly folding its thin pinnate leaves such that they point downwards, and carnivorous plants such as the Venus flytrap (*Dionaea muscipula*) produce specialized leaf structures that instantaneously snap shut when touched or landed upon by insects. In the Venus flytrap, touch is detected by cilia lining the inside of the specialized leaves, which generate an action potential that stimulates motor cells and causes movement to occur.

In plants, the mechanism responsible for adaptation is signal transduction. Adaptive responses include:

- Active foraging for light and nutrients. They do this by changing their architecture, eg branch growth and direction, physiology, and phenotype.

- Leaves and branches being positioned and oriented in response to a light source.

- Detecting soil volume and adapting growth accordingly, independently of nutrient availability.

- Defending against herbivores.

Aspects of Perception

Light

Many plant organs contain photo-sensitive compounds (phototropins, cryptochromes, and phytochromes), each of which reacts very specifically to certain wavelengths of light. These light sensors tell the plant if it is day or night, how long the day is, how much light is available, and from which direction the light is coming. Shoots grow towards light and roots usually grow away from light, responses known as phototropism and skototropism, respectively. They are brought about by light-sensitive pigments like phototropins and phytochromes and the plant hormone auxin.

The sunflower, a common heliotropic plant which perceives and reacts to sunlight by slow turning movement.

Many plants exhibit certain phenomena at specific times of the day; for example, certain flowers open only in the mornings. Plants keep track of the time of day with a circadian clock. This internal clock is synchronized with solar time every day using sunlight, temperature, and other cues, similar to the biological clocks present in other organisms. The internal clock coupled with the ability to perceive light also allows plants to measure the time of the day and so determine the season of the year. This is how many plants know when to flower. The seeds of many plants sprout only after they are exposed to light. This response is carried out by phytochrome signalling. Plants are also able to sense the quality of light and respond appropriately. For example, in low light conditions, plants produce more photosynthetic pigments. If the light is very bright or if the levels of harmful ultraviolet radiation increase, plants produce more of their protective pigments that act as sunscreens.

Gravity

To orient themselves correctly, plants must have an adequate sense of the direction of gravity's unidirectional pull. The subsequent response is known as gravitropism. In roots, this typically works as gravity is sensed and translated in the root tip, and subsequently roots grow towards gravity via elongation of the cells. In shoots, similar effects occur, but gravity is perceived and then growth occurs in the opposite direction, as the aboveground part of the plant experiences negative gravitropism.

At the root tip, amyloplasts containing starch granules fall in the direction of gravity. This weight activates secondary receptors, which signal to the plant the direction of the gravitational pull. After this occurs, auxin is redistributed through polar auxin transport and differential growth towards gravity begins. In the shoots, auxin redistribution occurs in a way to produce differential growth away from gravity.

For perception to occur, the plant often must be able to sense, perceive, and translate the direction of gravity. Without gravity, proper orientation will not occur and the plant will not effectively grow. The root will not be able to uptake nutrients or water, and the shoot will not grow towards the sky to maximize photosynthesis.

Plant Intelligence

Plants do not have brains or neuronal networks like animals do, at least in the traditional sense; however, reactions within signalling pathways may provide a biochemical basis for learning and memory in addition to computation and basic problem solving. Controversially, the brain is used as a metaphor in plant intelligence to provide an integrated view of signalling.

Plants respond to environmental stimuli by movement and changes in morphology. They communicate while actively competing for resources. In addition, plants accurately compute their circumstances, use sophisticated cost–benefit analysis, and take tightly controlled actions to mitigate and control diverse environmental stressors. Plants are also capable of discriminating between positive and negative experiences and of learning by registering memories from their past experiences. Plants use this information to adapt their behaviour in order to survive present and future challenges of their environments.

Plant physiology studies the role of signalling to integrate data obtained at the genetic, biochemical, cellular, and physiological levels, in order to understand plant development and behaviour. The neurobiological view sees plants as information-processing organisms with rather complex processes of communication occurring throughout the individual plant. It studies how environmental information is gathered, processed, integrated, and shared (sensory plant biology) to enable these adaptive and coordinated responses (plant behaviour); and how sensory perceptions and behavioural events are 'remembered' in order to allow predictions of future activities upon the basis of past experiences. Plants, it is claimed by some plant physiologists, are as sophisticated in behaviour as animals, but this sophistication has been masked by the time scales of plants' responses to stimuli, which are typically many orders of magnitude slower than those of animals.

It has been argued that although plants are capable of adaptation, it should not be called intelligence *per se*, as plant neurobiologists rely primarily on metaphors and analogies to argue that complex responses in plants can only be produced by intelligence. "A bacterium can monitor its environment and instigate developmental processes appropriate to the prevailing circumstances, but is that intelligence? Such simple adaptation behaviour might be bacterial intelligence but is clearly not animal intelligence". However, plant intelligence fits a definition of intelligence proposed by David Stenhouse in a book about evolution and animal intelligence, in which he describes it as "adaptively variable behaviour during the lifetime of the individual". Critics of the concept have also argued that a plant cannot have goals once it is past the developmental stage of seedling because, as a modular organism, each module seeks its own survival goals and the resulting organism-level behavior is not centrally controlled. This view, however, necessarily accommodates the possibility that a tree is a collection of individually intelligent modules cooperating, competing, and influencing each other to determine behavior in a bottom-up fashion. The development into a larger organism whose modules must deal with different environmental conditions and challenges is not universal across plant species, however, as smaller organisms might be subject to the same conditions across their bodies, at least, when the below and aboveground parts are considered separately. Moreover, the claim that central control of development is completely absent from plants is readily falsified by apical dominance.

The Italian botanist Federico Delpino wrote on the idea of plant intelligence in 1867. Charles Darwin studied movement in plants and in 1880 published a book, *The Power of Movement in Plants*. Darwin concludes:

> "It is hardly an exaggeration to say that the tip of the radicle thus endowed acts like the brain of one of the lower animals; the brain being situated within the anterior end of the body, receiving impressions from the sense-organs, and directing the several movements".

In philosophy, there are few studies of the implications of plant perception. Michael Marder put forth a phenomenology of plant life based on the physiology of plant perception. Paco Calvo Garzon offers a philosophical take on plant perception based on the cognitive sciences and the computational modeling of consciousness.

Comparison with Neurobiology

Plant sensory and response systems have been compared to the neurobiological processes of animals. Plant neurobiology concerns mostly the sensory adaptive behaviour of plants and plant electrophysiology. Indian scientist J. C. Bose is credited as the first person to research and talk about the neurobiology of plants. Many plant scientists and neuroscientists, however, view the term "plant neurobiology" as a misnomer, because plants do not have neurons.

The ideas behind plant neurobiology were criticised in a 2007 article published in *Trends in Plant Science* by Amedeo Alpi and 35 other scientists, including such eminent plant biologists as Gerd Jürgens, Ben Scheres, and Chris Sommerville. The breadth of fields of plant science represented by these researchers reflects the fact that the vast majority of the plant science research community rejects plant neurobiology as a legitimate notion. Their main arguments are that:

- "Plant neurobiology does not add to our understanding of plant physiology, plant cell biology or signaling".

- "There is no evidence for structures such as neurons, synapses or a brain in plants".

- The common occurrence of plasmodesmata in plants "poses a problem for signaling from an electrophysiological point of view", since extensive electrical coupling would preclude the need for any cell-to-cell transport of 'neurotransmitter-like' compounds.

The authors call for an end to "superficial analogies and questionable extrapolations" if the concept of "plant neurobiology" is to benefit the research community. Several responses to this criticism have attempted to clarify that the term "plant neurobiology" is a metaphor and that metaphors have proved useful on previous occasions. Plant ecophysiology describes this phenomenon.

Plant Stress

Plant stress is a state where a plant is growing in non-ideal growth conditions and has increased demands put on it. Plant stress refers to any unfavorable condition or substance that affects a plant's metabolism, reproduction, root development, or growth. Plant stress can come in different forms and durations. Some plant stressors are naturally occurring, like drought or wind, while others may be the result of human activity, like over irrigation or root disturbance.

As with humans, stresses can originate from the surrounding environment or, they can come from living organisms that can cause disease or damage.

Water Stress

One of the most important abiotic stresses affecting plants is water stress. A plant requires a certain amount of water for its optimal survival; too much water (flooding stress) can cause plant cells to swell and burst; whereas drought stress (too little water) can cause the plant to dry up, a condition called desiccation. Either condition can be deadly to the plant.

Temperature Stress

Temperature stresses can also wreak havoc on a plant. As with any living organism, a plant has an optimal temperature range at which it grows and performs best. If the temperature is too cold for the plant, it can lead to cold stress, also called chilling stress. Extreme forms of cold stress can lead to freezing stress. Cold temperatures can affect the amount and rate of uptake of water and nutrients, leading to cell desiccation and starvation. Under extremely cold conditions, the cell liquids can freeze outright, causing plant death.

Hot weather can affect plants adversely, too. Intense heat can cause plant cell proteins to break down, a process called denaturation. Cell walls and membranes can also "melt" under extremely high temperatures, and the permeability of the membranes is affected.

Other Abiotic Stresses

Other abiotic stresses are less obvious but can be equally as lethal. In the end, most abiotic stresses affect the plant cells in the same manner as do water stress and temperature stress. Wind stress can either directly damage the plant through sheer force; or, the wind can affect the transpiration of water through the leaf stomata and cause desiccation. Direct burning of plants through wildfires will cause the cell structure to break down through melting or denaturation.

In farming systems, the addition of agrochemicals such as fertilizers and pesticides, either in excess or in deficit, can also cause abiotic stress to the plant. The plant is affected by an imbalance of nutrition or via toxicity. High amounts of salt taken up by a plant can lead to cell desiccation, as elevated levels of salt outside a plant cell will cause water to leave the cell, a process called osmosis. Plant uptake of heavy metals can occur when plants grow in soils fertilized with improperly composted sewage sludge. High heavy metal content in plants can lead to complications with basic physiological and biochemical activities such as photosynthesis.

Biotic Stresses

Biotic stresses cause damage to plants via living organisms, including fungi, bacteria, insects, and weeds. Viruses, although they are not considered to be living organisms, also cause biotic stress to plants.

Fungi cause more diseases in plants than any other biotic stress factor. Over 8,000 fungal species are known to cause plant disease. On the other hand, only about 14 bacterial genera cause economically important diseases in plants, according to an Ohio State University Extension publication. Not many plant pathogenic viruses exist, but they are serious enough to cause nearly as much crop damage worldwide as fungi, according to published estimates. Microorganisms can cause plant wilt, leaf spots, root rot, or seed damage. Insects can cause severe physical damage to plants, including the leaves, stem, bark, and flowers. Insects can also act as a vector of viruses and bacteria from infected plants to healthy plants.

The method by which weeds, considered as unwanted and unprofitable plants, inhibit the growth of desirable plants such as crops or flowers is not by direct damage, but by competing with the desirable plants for space and nutrients. Because weeds grow quickly and produce an abundance of viable seed, they are often able to dominate environments more quickly than some desirable plants.

Plant Stress Measurement

Plant stress measurement is the quantification of environmental effects on plant health. When plants are subjected to less than ideal growing conditions, they are considered to be under stress. Stress factors can affect growth, survival and crop yields. Plant stress research looks at the response of plants to limitations and excesses of the main abiotic factors (light, temperature, water and nutrients), and of other stress factors that are important in particular situations (e.g. pests, pathogens, or pollutants). Plant stress measurement usually focuses on taking measurements from living plants. It can involve visual assessments of plant vitality, however, more recently the focus has moved to the use of instruments and protocols that reveal the response of particular processes within the plant (especially, photosynthesis, plant cell signalling and plant secondary metabolism)

- Determining the optimal conditions for plant growth, e.g. optimising water use in an agricultural system.

- Determining the climatic range of different species or subspecies.

- Determining which species or subspecies are resistant to a particular stress factor.

Instruments used to Measure Plant Stress

Measurements can be made from living plants using specialised equipment. Among the most commonly used instruments are those that measure parameters related to photosynthesis (chlorophyll content, chlorophyll fluorescence, gas exchange) or water use (porometer, pressure bomb). In addition to these general purpose instruments, researchers often design or adapt other instruments tailored to the specific stress response they are studying.

Photosynthesis Systems

Photosynthesis systems use infrared gas analyzers (IRGAS) for measuring photosynthesis. CO_2 concentration changes in leaf chambers are measured to provide carbon assimilation values for leaves or whole plants. Research has shown that the rate of photosynthesis is directly related to the amount of carbon assimilated by the plant. Measuring CO_2 in the air, before it enters the leaf chamber, and comparing it to air measured for CO_2 after it leaves the leaf chamber, provides this value using proven equations. These systems also use IRGAs, or solid state humidity sensors, for measuring H_2O changes in leaf chambers. This is done to measure leaf transpiration, and to correct CO_2 measurements. The light absorption spectrum for CO_2 and H_2O overlap somewhat, therefore, a correction is necessary for reliable CO_2 measuring results. The critical measurement for most plant stress measurements is designated by "A" or carbon assimilation rate. When a plant is under stress, less carbon is assimilated. CO_2 IRGAs are capable of measuring to approximately +/- 1 µmol or 1ppm of CO_2.

Because these systems are effective in measuring carbon assimilation and transpiration at low rates, as found in stressed plants, they are often used as the standard to compare to other types of instruments. Photosynthesis instruments come in field portable and laboratory versions. They are also designed to measure ambient environmental conditions, and some systems offer variable microclimate control of the measuring chamber.

Microclimate control systems allow adjustment of the measuring chamber temperature, CO_2 level, light level, and humidity level for more detailed investigation.

The combination of these systems with fluorometers, can be especially effective for some types of stress, and can be diagnostic, e.g. in the study of cold stress and drought stress.

Chlorophyll Fluorometers

Chlorophyll fluorescence emitted from plant leaves gives an insight into the health of the photosynthetic systems within the leaf. Chlorophyll fluorometers are designed to measure variable fluorescence of photosystem II. This variable fluorescence can be used to measure the level of plant stress. The most commonly used protocols include those aimed at measuring the photosynthetic efficiency of photosystem II, both in the light (ΔF/Fm') and in a dark-adapted state (Fv/Fm). Chlorophyll fluorometers are, for the most part, less expensive tools than photosynthesis systems, they also have a faster measurement time and tend to be more portable. For these reasons they have become one of the most important tools for field measurements of plant stress.

Fv/Fm

Fv/Fm tests whether or not plant stress affects photosystem II in a dark adapted state. Fv/Fm is the most used chlorophyll fluorescence measuring parameter in the world. "The majority of fluorescence measurements are now made using modulated fluorometers with the leaf poised in a known state".

Light that is absorbed by a leaf follows three competitive pathways. It may be used in photochemistry to produce ATP and NADPH used in photosynthesis, it can be re-emitted as fluorescence, or dissipated as heat. The Fv/Fm test is designed to allow the maximum amount of the light energy to take the fluorescence pathway. It compares the dark-adapted leaf pre-photosynthetic fluorescent state, called minimum fluorescence, or Fo, to maximum fluorescence called Fm. In maximum fluorescence, the maximum number of reaction centers have been reduced or closed by a saturating light source. In general, the greater the plant stress, the fewer open reaction centers available, and the Fv/Fm ratio is lowered. Fv/Fm is a measuring protocol that works for many types of plant stress.

In Fv/Fm measurements, after dark adaption, minimum fluorescence is measured, using a modulated light source. This is a measurement of antennae fluorescence using a modulated light intensity that is too low to drive photosynthesis. Next, an intense light flash, or saturation pulse, of a limited duration, is used, to expose the sample, and close all available reaction centers. With all available reaction centers closed, or chemically reduced, maximum fluorescence is measured. The difference between maximum fluorescence and minimum fluorescence is Fv, or variable fluorescence. Fv/Fm is a normalize ratio created by dividing variable fluorescence by maximum fluorescence.

It is a measurement ratio that represents the maximum potential quantum efficiency of Photosystem II if all capable reaction centers were open. An Fv/Fm value in the range of 0.79 to 0.84 is the approximate optimal value for many plant species, with lowered values indicating plant stress (Maxwell K., Johnson G. N. 2000), (Kitajima and Butler, 1975). Fv/Fm is a fast test that usually takes a few seconds. It was developed in and around 1975 by Kitajima and Butler. Dark adaptation times vary from about fifteen minutes to overnight. Some researchers will only use pre-dawn values.

Y(II) or ΔF/Fm' and ETR

Y(II) is a measuring protocol that was developed by Bernard Genty with the first publications in 1989 and 1990. It is a light adapted test that allows one to measure plant stress while the plant is undergoing the photosynthetic process at steady-state photosynthesis lighting conditions. Like FvFm, Y(II) represents a measurement ratio of plant efficiency, but in this case, it is an indication of the amount of energy used in photochemistry by photosystem II under steady-state photosynthetic lighting conditions. For most types of plant stress, Y(II) correlates to plant carbon assimilation in a linear fashion in C_4 plants. In C_3 plants, most types of plant stress correlate to carbon assimilation in a curve-linear fashion. According to Maxwell and Johnson, it takes between fifteen and twenty minutes for a plant to reach steady-state photosynthesis at a specific light level. In the field, plants in full sunlight, and not under canopy, or partly cloudy conditions, are considered to be at steady state. In this test, light irradiation levels and leaf temperature must be controlled or measured, because while the Y(II) parameter levels vary with most types of plant stress, it also varies with light level and temperature. Y(II) values will be higher at lower light levels than at higher light levels. Y(II) has the advantage that it is more sensitive to a larger number of plant stress types than Fv/Fm.

ETR, or electron transport rate, is also a light-adapted parameter that is directly related to Y(II) by the equation, ETR = Y(II) × PAR × 0.84 × 0.5. By multiplying Y(II) by the irradiation light level in the PAR range (400 nm to 700 nm) in μmols, multiplied by the average ratio of light absorbed by the leaf 0.84, and multiplied by the average ratio of PSII reaction centers to PSI reaction centers, 0.50, relative ETR measurement is achieved.

Relative ETR values are valuable for stress measurements when comparing one plant to another, as long as the plants to be compared have similar light absorption characteristics. Leaf absorption characteristics can vary by water content, age, and other factors. If absorption differences are a concern, absorption can be measured with the use of an integrating sphere. For more accurate ETR values, the leaf absorption value and the ratio of PSII reaction centers to PSI reaction centers can be included in the equation. If different leaf absorption ratios are an issue, or they are an unwanted variable, then using Y(II) instead of ETR, may be the best choice. Four electrons must be transported for every CO_2 molecule assimilated, or O_2 molecule evolved, differences from gas exchange measurements, especially in C_3 plants, can occur under conditions that promote photorespiration, cyclic electron transport, and nitrate reduction.

Quenching Measurements

Quenching measurements have been traditionally used for light stress, and heat stress measurements. In addition, they have been used to study plant photoprotective mechanisms, state transitions, plant photoinhibition, and the distribution of light energy in plants. While they can be used for many types of plant stress measurement, the time required, and the additional expense required for this capability, limit their use. These tests commonly require overnight dark adaptation, and fifteen to twenty minutes in lighted conditions to reach steady state photosynthesis before measurement.

Puddle Model and Lake Model Quenching Parameters

"Understanding of the organization of plant antennae, or plant light collection structures, and reaction centers, where the photosynthetic light reaction actually takes place, has changed over the years. It is now understood that a single antennae does not link only to a single reaction center, as was previously described in the puddle model. Current evidence indicates that reaction centers are connected with shared antennae in terrestrial plants". As a result, the parameters used to provide reliable measurements have changed to represent the newer understanding of this relationship. The model that represents the newer understanding of the antennae - reaction center relationship is called the lake model.

Lake model parameters were provided by Dave Kramer in 2004. Since then, Luke Hendrickson has provided simplified lake model parameters that allow the resurrection of the parameter NPQ, from the puddle model, back into the lake model. This is valuable because there have been so many scientific papers that have used NPQ for plant stress measurement, as compared to papers that have used lake model parameters.

OJIP or OJIDP

OJIP or OJIDP is a dark adapted chlorophyll fluorescence technique that is used for plant stress measurement. It has been found that by using a high time resolution scale, the rise to maximum fluorescence from minimum fluorescence has intermediate peaks and dips, designated by the OJID and P nomenclature. Over the years, there have been multiple theories of what the rise, time scale, peaks and dips mean. In addition, there is more than one school as to how this information should be used for plant stress testing (Strasser 2004), (Vredenburg 2004, 2009, 2011). Like Fv/Fm, and the other protocols, the research shows that OJIP works better for some types of plant stress than it does for others.

Choosing the Best Chlorophyll Fluorescence Protocol and Parameter

When choosing the correct protocol, and measuring parameter, for a specific type of plant stress, it is important to understand the limitations of the instrument, and the protocol used. For example, it was found that when measuring Oak leaves, a photosynthesis system could detect heat stress at 30 °C and above, Y(II) could detect heat stress at 35 °C and above, NPQ could detect heat stress at 35 °C and above, and Fv/Fm could only detect heat stress at 45 °C and above. OJIP was found to detect heat stress at 44 °C and above on samples tested.

The relationship between carbon assimilation measurements made by photosynthesis systems of the dark Calvin cycle, and measurements of variable fluorescence of photosystem II (PSII), made by chlorophyll fluorometers of the light reaction, are not always straightforward. For this reason, choosing the correct chlorophyll fluorescence protocol can also be different for C_3 and C_4 plants. It has been found, for example, that Y(II) and ETR are good tests for drought stress in C_4 plants, but a special assay is required to measure drought stress in most C_3 plants at usable levels. In C_3 plants, photorespiration, and the Mehler reaction, are thought to be a principal cause.

Chlorophyll Content Meters

These are instruments that use light transmission through a leaf, at two wavelengths, to determine the greenness and thickness of leaves. Transmission in the infrared range provides a measurement related to leaf thickness, and a wavelength in the red light range is used to determine greenness. The ratio of the transmission of the two wavelengths provides a chlorophyll content index that is referred to as CCI or alternatively as a SPAD index. CCI is a linear scale, and SPAD is a logarithmic scale. These instruments and scales have been shown to correlate to chlorophyll chemical tests for chlorophyll content except at very high levels.

Chlorophyll content meters are commonly used for nutrient plant stress measurement, that includes nitrogen stress, and sulfur stress. Because research has shown, that if used correctly, chlorophyll content meters are reliable for nitrogen management work, these meters are often the instruments of choice for crop fertilizer management because they are relatively inexpensive. Research has demonstrated that by comparing well fertilized plants to test plants, the ratio of the chlorophyll content index of test plants, divided by the chlorophyll content index of well fertilized plants, will provide a ratio that is an indication of when fertilization should occur, and how much should be used. It is common to use a well fertilized stand of crops in a specific field and sometimes in different areas of the same field, as the fertilization reference, due to differences from field to field and within a field. The research done to date uses either ten and thirty measurements on test and well fertilized crops, to provide average values. Studies have been done for corn and wheat. One paper suggests that when the ratio drops below 95%, it is time to fertigate. The amounts of fertilizer are also recommended.

Crop consultants also use these tools for fertilizer recommendations. However, because strict scientific protocols are more time consuming and more expensive, consultants sometimes use well-fertilized plants located in low-lying areas as the standard well-fertilized plants. They typically also use fewer measurements. The evidence for this approach involves anecdotal discussions with crop consultants. Chlorophyll content meters are sensitive to both nitrogen and sulfur stress at usable levels. Chlorophyll fluorometers require a special assay, involving a high actinic light level in combination with nitrogen stress, to measure nitrogen stress at usable levels. In addition, chlorophyll fluorometers will only detect sulfur stress at starvation levels. For best results, chlorophyll content measurements should be made when water deficits are not present. Photosynthesis systems will detect both nitrogen and sulfur stress.

References

- Introduction to california plant life, robert ornduff, phyllis m. Faber, todd keeler-wolf, california natural history guides no. 69, university of california press, ltd., 2003, isbn 978-0-520-23704-9

- Scott, peter (2008). Physiology and behaviour of plants. West sussex, england: john wiley and sons, ltd. Isbn 978-0-470-85024-4

- Michmizos d, chilioti z (january 2019). "a roadmap towards a functional paradigm for learning & memory in plants". Journal of plant physiology. 232 (1): 209–215. Doi:10.1016/j.jplph.2018.11.002. Pmid 30537608

- Baluška f, volkmann d, mancuso s (2006). Communication in plants: neuronal aspects of plant life. Springer verlag. Isbn 978-3-540-28475-8

- Fromm, jörg; lautner, silke (march 2007). "electrical signals and their physiological significance in plants: electrical signals in plants". Plant, cell & environment. 30 (3): 249–257. Doi:10.1111/j.1365-3040.2006.01614.x. Pmid 17263772

- Plant-stresses-abiotic-and-biotic-stresses-419223: thoughtco.com, Retrieved 25 February, 2019

Human Ecology 4

- **Cold and Heat Adaptations in Humans**
- **Coupled Human–environment System**
- **Urban Ecology**
- **New Urbanism**
- **Anthropocene**

Human ecology refers to the study of relationship between human and their social, natural and built environments. Urban ecology, new urbanism, anthropocene, etc. are some of the aspects that fall under its domain. This chapter closely examines these key concepts of human ecology to provide an extensive understanding of the subject.

Human ecology refers to man's collective interaction with his environment. Influenced by the work of biologists on the interaction of organisms within their environments, social scientists undertook to study human groups in a similar way.

Thus, ecology in the social sciences is the study of the ways in which the social structure adapts to the quality of natural resources and to the existence of other human groups. When this study is limited to the development and variation of cultural properties, it is called cultural ecology.

Human ecology views the biological, environmental, demographic, and technical conditions of the life of any people as an interrelated series of determinants of form and function in human cultures and social systems.

It recognizes that group behaviour is dependent upon resources and associated skills and upon a body of emotionally charged beliefs; these together give rise to a system of social structures.

Cold and Heat Adaptations in Humans

Cold and heat adaptations in humans are a part of the broad adaptability of *Homo sapiens*. Adaptations in humans can be physiological, genetic, or cultural, which allow people to live in a wide variety of climates. There has been a great deal of research done on developmental adjustment, acclimatization, and cultural practices, but less research on genetic adaptations to cold and heat temperatures.

The human body always works to remain in homeostasis. One form of homeostasis is thermoregulation. Body temperature varies in every individual, but the average internal temperature is 37.0 °C (98.6 °F). Stress from extreme external temperature can cause the human body to shut down. Hypothermia can set in when the core temperature drops to 35 °C (95 °F). Hyperthermia can set in when the core body temperature rises above 37.5-38.3 °C (99.5-100.9 °F). These temperatures commonly result in mortality. Humans have adapted to, such as the use of clothing and shelter and they have adapted to go to the washroom and how to make uful stuff for living.

Origin of Cold and Heat Adaptations

Modern humans emerged from Africa approximately 40,000 years ago during a period of unstable climate, leading to a variety of new traits among the population. When modern humans spread into Europe, they outcompeted Neanderthals. Researchers hypothesize that this suggests early modern humans were more evolutionarily fit to live in various climates. This is supported in the variability selection hypothesis proposed by Richard Potts, which says that human adaptability came from environmental change over the long term.

Ecogeographic Rules

Bergmann's rule states that endothermic animal subspecies living in colder climates have larger bodies than that of the subspecies living in warmer climates. Individuals with larger bodies are better suited for colder climates because larger bodies produce more heat due to having more cells, and have a smaller surface area to volume ratio compared to smaller individuals, which reduces heat loss. A study by Frederick Foster and Mark Collard found that Bergmann's rule can be applied to humans when the latitude and temperature between groups differ widely.

Allen's rule is a biological rule that says the limbs of endotherms are shorter in cold climates and longer in hot climates. Limb length affects the body's surface area, which helps with thermoregulation. Shorter limbs help to conserve heat, while longer limbs help to dissipate heat. Marshall T. Newman argues that this can be observed in Eskimo, who have shorter limbs than other people and are laterally built.

Physiological Adaptations

Origins of heat and cold adaptations can be explained by climatic adaptation. Ambient air temperature affects how much energy investment the human body must make. The temperature that requires the least amount of energy investment is 21 °C (69.8 °F). The body controls its temperature through the hypothalamus. Thermoreceptors in the skin send signals to the hypothalamus, which indicate when vasodilation and vasoconstriction should occur.

Cold

The human body has two methods of thermogenesis, which produces heat to raise the core body temperature. The first is shivering, which occurs in an unclothed person when the ambient air temperature is under 25 °C (77 °F). It is limited by the amount of glycogen available in the body. The second is non-shivering, which occurs in brown adipose tissue.

Population studies have shown that the San tribe of Southern Africa and the Sandawe of Eastern Africa have reduced shivering thermogenesis in the cold, and poor cold induced vasodilation in fingers and toes compared to that of Caucasians.

Heat

The only mechanism the human body has to cool itself is by sweat evaporation. Sweating occurs when the ambient air temperatures is above 28 °C (82 °F) and the body fails to return to the normal internal temperature. The evaporation of the sweat helps cool the blood beneath the skin. It is limited by the amount of water available in the body, which can cause dehydration.

Humans adapted to heat early on. In Africa, the climate selected for traits that helped us stay cool. Also, we had physiological mechanisms that reduced the rate of metabolism and that modified the sensitivity of sweat glands to provide an adequate amount for cooldown without the individual becoming dehydrated.

There are two types of heat the body is adapted to, humid heat and dry heat, but the body has adapted to both in the same way. Humid heat is characterized by warmer temperatures with a high amount of water vapor in the air. Humid heat is dangerous as the moisture in the air prevents the evaporation of sweat. Dry heat is characterized by warmer temperatures with little to no water vapor in the air, such as desert conditions. Dry heat is also very dangerous as sweat will tend to evaporate extremely quickly, causing dehydration. Both humid heat and dry heat favor individuals with less fat and slightly lower body temperatures.

Acclimatization

When humans are exposed to certain climates for extended periods of time, physiological changes occur to help the individual adapt to hot or cold climates. This helps the body conserve energy.

Cold

The Inuit have more blood flowing into their extremities, and at a hotter temperature, than people living in warmer climates. A 1960 study on the Alacaluf Indians shows that they have a resting metabolic rate 150 to 200 percent higher than the white controls used. Lapps do not have an increase in metabolic rate when sleeping, unlike non-acclimated people. Australian aborigines undergo a similar process, where the body cools but the metabolic rate does not increase.

Heat

Humans in Central Africa have been living in similar tropical climates for at least 40,000 years, which means that they have similar thermoregulatory systems.

A study done on the Bantus of South Africa showed that Bantus have a lower sweat rate than that of acclimated and nonacclimated whites. A similar study done on Australian aborigines produced similar results, with aborigines having a much lower sweat rate than whites.

Culture

Social adaptations enabled early modern humans to occupy environments with temperatures that were drastically different from that of Africa. Culture enabled humans to expand their range to areas that would otherwise be uninhabitable.

Cold

Humans have been able to occupy areas of extreme cold through clothing, buildings, and manipulation of fire. Furnaces have further enabled the occupation of cold environments.

Australian aborigines only wear genital coverings for clothes, but studies have shown that the warmth from the fires they build is enough to keep the body from fighting heat loss through shivering. Eskimos use well-insulated houses that are designed to transfer heat from an energy source to the living area, which means that the average indoor temperature for coastal Eskimos is 10 to 20 °C (50-68 °F).

Heat

Humans inhabit hot climates, both dry and humid, and have done so for thousands of years. Selective use of clothing and technological inventions such as air conditioning allows humans to thrive in hot climates.

One example is the Chaamba Arabs, who live in the Sahara Desert. They wear clothing that traps air in between skin and the clothes, preventing the high ambient air temperature from reaching the skin.

Genetic Adaptations

There has been very little research done in the genetics behind adaptations to heat and cold stress. Data suggests that certain parts of the human genome have only been selected for recently. Research on gene-culture interaction has been successful in linking agriculture and lactose tolerance. However, most evidence of links between culture and selection has not been proven.

Coupled Human–environment System

Coupled human and natural systems (CHANS; or coupled social-ecological systems or coupled human-environment systems) are integrated and complex systems in which humans and nature interact with one another. Wildlife are important components of CHANS because they interact with humans in numerous complex ways in today's increasingly human-influenced world. Globally, the continuing conversion of natural ecosystems to areas used intensively by humans has greatly reduced wildlife habitat, leading to an "extinction crisis". The disappearance of wildlife and their habitats entails the degradation of life-sustaining ecosystem services such as the availability of medicines, control of pests and diseases, and provision of clean water and air. Moreover, because people worldwide value nature for numerous reasons (e.g., aesthetic, cultural, religious, economic, educational), the loss of wildlife and their habitats diminishes humans' quality of life.

Given these challenges, an integrated CHANS approach for understanding human-wildlife interactions is of utmost importance. Although interactions between people and wildlife have been examined for some time, most studies are compartmentalized within disciplines. There is little knowledge on how people and wildlife are interlinked, across space and through time, from combined social and environmental perspectives, together with the mechanisms that may weaken or strengthen those linkages. Thus, to reach broad, generalizable insights about wildlife dynamics, findings from sites with different ecological, socioeconomic, political, demographic, and cultural settings need to be synthesized. Such cross-site syntheses will have significant scientific value and facilitate knowledge exchange among multiple stakeholders, including local residents, managers of natural resources, policy makers, tourists, and researchers. This is critical for developing an array of policies and interventions that improve human wellbeing while sustaining wildlife populations and their habitats.

The Coupled Human and Natural Systems Approach

Inherently integrative in nature, the CHANS approach brings together theoretical and analytical techniques from diverse disciplines, including those from ecological and social sciences, to understand the nuances of such complex systems. The CHANS approach is thus well suited for understanding wildlife dynamics in human-influenced landscapes. First, by transcending a single discipline, the approach can account for the patterns and processes that link people and their activities with wildlife and their habitats. Second, rather than focusing on unidirectional relationships, the approach can identify key relationships and feedbacks between people and wildlife. Third, the approach facilitates understanding of cross-scale (e.g., spatial, temporal, and organizational) interactions between people and wildlife. Thus, the CHANS approach can better clarify relationships between people and wildlife and consequently help to prevent further habitat loss and wildlife population decline in the face of synergistic and increasingly complex threats (e.g., overexploitation of natural resources, climate change) while simultaneously responding to growing human aspirations for improved quality of life.

We conceptualize each of the focal sites as a coupled system consisting of two main subsystems, the human subsystem and the natural subsystem, whose detailed operation is made evident through appropriate disciplinary analyses. For our purposes, the human subsystem comprises communities and local residents, and the natural subsystem comprises wildlife and the land cover characterizing their habitat. Telecouplings (i.e., socioeconomic and environmental interactions over distances) link the focal coupled system to other distant coupled systems. The characteristics of each of these system components are interrelated and influence the characteristics of the other system components. For example, collection of fuelwood and grasses by local residents can change land cover composition and structure and disrupt the spatial and temporal distribution of wildlife. Community organizations such as forest user group committees can encourage reforestation, which in turn provides more income for local residents. Networks of people in the community connected through "weak ties" can shape and be shaped by policies and norms that influence many aspects of the daily lives of local residents. Wildlife can change forest characteristics through browsing and might decrease household income by eating crops and livestock. The interactions within and among each of these components influence and are influenced by telecouplings such as tourism and migration, among many others. For example, labor demands by urban centers generate rural-urban migration, which can reduce the proportion of young people living in a rural area. Fewer young people collecting natural resources from nearby forests can slow rates of wildlife habitat degradation.

Worldviews differ markedly with respect to the way people understand the world, and particularly, what it is to be human and the role humans play in that world. The CHANS approach is intended neither to perpetuate one worldview (e.g., people separate from nature) over another, nor to ignore the perceptions and knowledge of certain groups of people. Rather, it is intended to serve as a pragmatic, heuristic tool for analyzing interrelationships between people and the environment, seeking to reunite the scientific traditions focusing on particular subsystems. The CHANS framework emphasizes that the human and natural components are coupled rather than separate. Furthermore, it emphasizes feedbacks between the components. Collaboration among a range of stakeholders helps CHANS projects build alternative hypotheses and understandings of complex issues.

Urban Ecology

Urban ecology is the scientific study of the relation of living organisms with each other and their surroundings in the context of an urban environment. The urban environment refers to environments dominated by high-density residential and commercial buildings, paved surfaces, and other urban-related factors that create a unique landscape dissimilar to most previously studied environments in the field of ecology.

Urban ecology is a recent field of study compared to ecology as a whole. The methods and studies of urban ecology are similar to and comprise a subset of ecology. The study of urban ecology carries increasing importance because more than 50% of the world's population today lives in urban areas. At the same time, it is estimated that within the next forty years, two-thirds of the world's population will be living in expanding urban centers. The ecological processes in the urban environment are comparable to those outside the urban context. However, the types of urban habitats and the species that inhabit them are poorly documented. Often, explanations for phenomena examined in the urban setting as well as predicting changes because of urbanization are the center for scientific research.

Methods

Since urban ecology is a subfield of ecology, many of the techniques are similar to that of ecology. Ecological study techniques have been developed over centuries, but many of the techniques use for urban ecology are more recently developed. Methods used for studying urban ecology involve chemical and biochemical techniques, temperature recording, heat mapping remote sensing, and long-term ecological research sites.

Chemical and Biochemical Techniques

Chemical techniques may be used to determine pollutant concentrations and their effects. Tests can be as simple as dipping a manufactured test strip, as in the case of pH testing, or be more complex, as in the case of examining the spatial and temporal variation of heavy metal contamination due to industrial runoff. In that particular study, livers of birds from many regions of the North Sea were ground up and mercury was extracted. Additionally, mercury bound in feathers was extracted from both live birds and from museum specimens to test for mercury levels across many decades. Through these two different measurements, researchers were able to make a complex picture of the spread of mercury due to industrial runoff both spatially and temporally.

Other chemical techniques include tests for nitrates, phosphates, sulfates, etc. which are commonly associated with urban pollutants such as fertilizer and industrial by-products. These biochemical fluxes are studied in the atmosphere (e.g. greenhouse gasses), aquatic ecosystems and soil vegetation. Broad reaching effects of these bio-chemical fluxes can be seen in various aspects of both the urban and surrounding rural ecosystems.

Temperature Data and Heat Mapping

Temperature data can be used for various kinds of studies. An important aspect of temperature data is the ability to correlate temperature with various factors that may be affecting or occurring in the environment. Oftentimes, temperature data is collected long-term by the Office of Oceanic and Atmospheric Research (OAR), and made available to the scientific community through the National Oceanic and Atmospheric Administration (NOAA). Data can be overlaid with maps of terrain, urban features, and other spatial areas to create heat maps. These heat maps can be used to view trends and distribution over time and space.

Remote Sensing

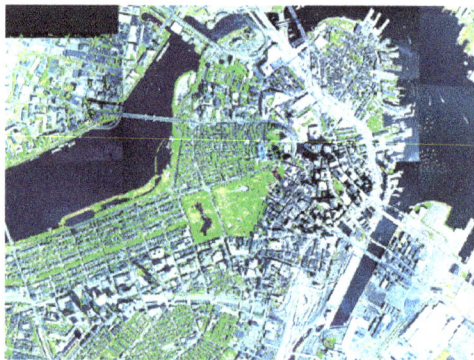

Remote sensing allows collection of data using satellites.
This map shows urban tree canopy.

Remote sensing is the technique in which data is collected from distant locations through the use of satellite imaging, radar, and aerial photographs. In urban ecology, remote sensing is used to collect data about terrain, weather patterns, light, and vegetation. One application of remote sensing for urban ecology is to detect the productivity of an area by measuring the photosynthetic wavelengths of emitted light. Satellite images can also be used to detect differences in temperature and landscape diversity to detect the effects of urbanization.

LTERs and Long-term Data Sets

Long-term ecological research (LTER) sites are research sites funded by the government that have collected reliable long-term data over an extended period of time in order to identify long-term climatic or ecological trends. These sites provide long-term temporal and spatial data such as average temperature, rainfall and other ecological processes. The main purpose of LTERs for urban ecologists is the collection of vast amounts of data over long periods of time. These long-term data sets can then be analyzed to find trends relating to the effects of the urban environment on various ecological processes, such as species diversity and abundance over time. Another example is the examination of temperature trends that are accompanied with the growth of urban centers.

Urban Effects on the Environment

Humans are the driving force behind urban ecology and influence the environment in a variety of ways, such as modifying land surfaces and waterways, introducing foreign species, and altering biogeochemical cycles. Some of these effects are more apparent, such as the reversal of the Chicago River to accommodate the growing pollution levels and trade on the river. Other effects can be more gradual such as the change in global climate due to urbanization.

Modification of Land and Waterways

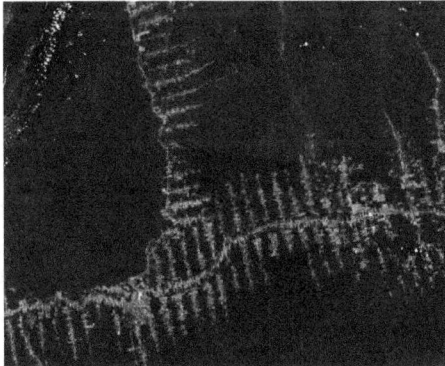

Deforestation in the Amazon rainforest. The "fishbone pattern" is a result of the roads in the forest created by loggers.

Humans place high demand on land not only to build urban centers, but also to build surrounding suburban areas for housing. Land is also allocated for agriculture to sustain the growing population of the city. Expanding cities and suburban areas necessitate corresponding deforestation to meet the land-use and resource requirements of urbanization. Key examples of this are Deforestation in the United States and Brazil.

Along with manipulation of land to suit human needs, natural water resources such as rivers and streams are also modified in urban establishments. Modification can come in the form of dams, artificial canals, and even the reversal of rivers. Reversing the flow of the Chicago River is a major example of urban environmental modification. Urban areas in natural desert settings often bring in water from far areas to maintain the human population and will likely have effects on the local desert climate. Modification of aquatic systems in urban areas also results in decreased stream diversity and increased pollution.

Trade, Shipping and Spread of Invasive Species

A ship navigates through the Firth of Clyde,
potentially carrying invasive species.

Invasive kudzu vines growing on trees.

Both local shipping and long-distance trade are required to meet the resource demands important in maintaining urban areas. Carbon dioxide emissions from the transport of goods also contribute to accumulating greenhouse gases and nutrient deposits in the soil and air of urban environments. In addition, shipping facilitates the unintentional spread of living organisms, and introduces them to environments that they would not naturally inhabit. Introduced or alien species are populations of organisms living in a range in which they did not naturally evolve due to intentional or inadvertent human activity. Increased transportation between urban centers furthers the incidental movement of animal and plant species. Alien species often have no natural predators and pose a substantial threat to the dynamics of existing ecological populations in the new environment where they are introduced. Such invasive species are numerous and include house sparrows, ring-necked pheasants, European starlings, brown rats, Asian carp, American bullfrogs, emerald ash borer, kudzu vines, and zebra mussels among numerous others, most notably domesticated animals. In Australia, it has been found that removing Lantana (*L. camara,* an alien species) from urban greenspaces can surprisingly have negative impacts on bird diversity locally, as it provides refugia for species like the superb fairy (Malurus cyaneus) and silvereye (Zosterops lateralis), in the absence of native plant equivalents. Although, there seems to be a density threshold in which too much Lantana (thus homogeneity in vegetation cover) can lead to a decrease in bird species richness or abundance.

Human Effects on Biogeochemical Pathways

Urbanization results in a large demand for chemical use by industry, construction, agriculture, and energy providing services. Such demands have a substantial impact on biogeochemical cycles, resulting in phenomena such as acid rain, eutrophication, and global warming. Furthermore, natural biogeochemical cycles in the urban environment can be impeded due to impermeable surfaces that prevent nutrients from returning to the soil, water, and atmosphere.

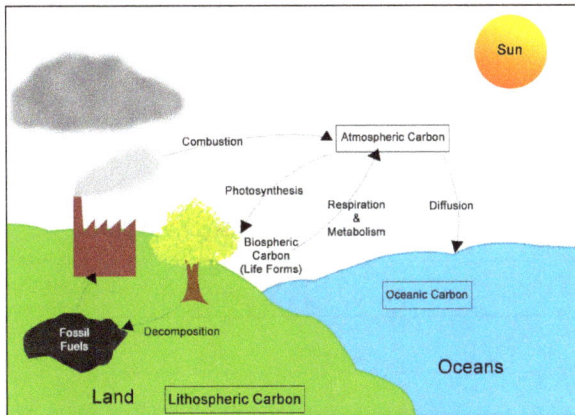

Graphical representation of the carbon cycle.

Demand for fertilizers to meet agricultural needs exerted by expanding urban centers can alter chemical composition of soil. Such effects often result in abnormally high concentrations of compounds including sulfur, phosphorus, nitrogen, and heavy metals. In addition, nitrogen and phosphorus used in fertilizers have caused severe problems in the form of agricultural runoff, which alters the concentration of these compounds in local rivers and streams, often resulting in adverse effects on native species. A well-known effect of agricultural runoff is the phenomenon of eutrophication. When the fertilizer chemicals from agricultural runoff reach the ocean, an algal bloom results, then rapidly dies off. The dead algae biomass is decomposed by bacteria that also consume large quantities of oxygen, which they obtain from the water, creating a "dead zone" without oxygen for fish or other organisms. A classic example is the dead zone in the Gulf of Mexico due to agricultural runoff into the Mississippi River.

Just as pollutants and alterations in the biogeochemical cycle alter river and ocean ecosystems, they exert likewise effects in the air. Smog stems from the accumulation of chemicals and pollution and often manifests in urban settings, which has a great impact on local plants and animals. Because urban centers are often considered point sources for pollution, unsurprisingly local plants have adapted to withstand such conditions.

Urban Effects on Climate

Urban environments and outlying areas have been found to exhibit unique local temperatures, precipitation, and other characteristic activity due to a variety of factors such as pollution and altered geochemical cycles. Some examples of the urban effects on climate are urban heat island, oasis effect, greenhouse gases, and acid rain. This further stirs the debate as to whether urban areas should be considered a unique biome. Despite common trends among all urban centers, the surrounding local environment heavily influences much of the climate. One such example of regional differences can be seen through the urban heat island and oasis effect.

Urban Heat Island Effect

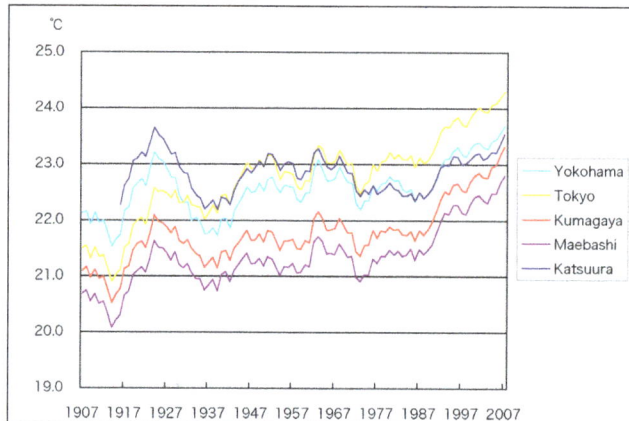

Graphical representation of the rising temperature in Kanto, Japan due to urban heat island.

The urban heat island is a phenomenon in which central regions of urban centers exhibit higher mean temperatures than surrounding urban areas. Much of this effect can be attributed to low city albedo, the reflecting power of a surface, and the increased surface area of buildings to absorb solar radiation. Concrete, cement, and metal surfaces in urban areas tend to absorb heat energy rather than reflect it, contributing to higher urban temperatures. Brazel found that the urban heat island effect demonstrates a positive correlation with population density in the city of Baltimore. The heat island effect has corresponding ecological consequences on resident species. However, this effect has only been seen in temperate climates.

Greenhouse Gases

Greenhouse gas emissions include those of carbon dioxide and methane from the combustion of fossil fuels to supply energy needed by vast urban metropolises. Other greenhouse gases include water vapor, and nitrous oxide. Increases in greenhouse gases due to urban transport, construction, industry and other demands have been correlated strongly with increase in temperature. Sources of methane are agricultural dairy cows and landfills.

Acid Rain and Pollution

Smokestacks from a wartime production plant releasing pollutants into the atmosphere.

Processes related to urban areas result in the emission of numerous pollutants, which change corresponding nutrient cycles of carbon, sulfur, nitrogen, and other elements. Ecosystems in and around the urban center are especially influenced by these point sources of pollution. High sulfur dioxide concentrations resulting from the industrial demands of urbanization cause rainwater to become more acidic. Such an effect has been found to have a significant influence on locally affected populations, especially in aquatic environments. Wastes from urban centers, especially large urban centers in developed nations, can drive biogeochemical cycles on a global scale.

Urban Environment as an Anthropogenic Biome

The urban environment has been classified as an anthropogenic biome, which is characterized by the predominance of certain species and climate trends such as urban heat island across many urban areas. Examples of species characteristic of many urban environments include, cats, dogs, mosquitoes, rats, flies, and pigeons, which are all generalists. Many of these are dependent on human activity and have adapted accordingly to the niche created by urban centers.

Biodiversity and Urbanization

Research thus far indicates that, on a small scale, urbanization often increases the biodiversity of non-native species while reducing that of native species. This normally results in an overall reduction in species richness and increase in total biomass and species abundance. Urbanization also reduces diversity on a large scale.

Urban stream syndrome is a consistently observed trait of urbanization character-ized by high nutrient and contaminant concentration, altered stream morphology, in-creased dominance of dominant species, and decreased biodiversity The two primary causes of urban stream syndrome are storm water runoff and wastewater treatment plant effluent.

Changes in Diversity

Diversity is normally reduced at intermediate-low levels of urbanization but is always reduced at high levels of urbanization. These effects have been observed in vertebrates and invertebrates while plant species tend to increase with intermediate-low levels of urbanization but these general trends do not apply to all organisms within those groups. For example, McKinney's review did not include the effects of urbanization on fishes and of the 58 studies on invertebrates, 52 included insects while only 10 includ-ed spiders. There is also a geographical bias as most of the studies either took place in North America or Europe.

The effects of urbanization also depend on the type and range of resources used by the organism. Generalist species, those that use a wide range of resources and can thrive under a large range of living conditions, are likely survive in uniform environments. Specialist species, those that use a narrow range of resources and can only cope with a narrow range of living conditions, are unlikely to cope with uniform environments. There will likely be a variable effect on these two groups of organisms as urbanization alters habitat uniformity. Surprisingly, endangered plant species have been reported to occur throughout a wide range of urban ecosystems, many of them being novel ecosys-tems.

Cause of Diversity Change

The urban environment can decrease diversity through habitat removal and species homogenization - the increasing similarity between two previously distinct biological communities. Habitat degradation and habitat fragmentation reduces the amount of suitable habitat by urban development and separates suitable patches by inhospitable terrain such as roads, neighborhoods, and open parks. Although this replacement of suitable habitat with unsuitable habitat will result in extinctions of native species, some shelter may be artificially created and promote the survival of non-native species (e.g. house sparrow and house mice nests). Urbanization promotes species homogenization through the extinction of native endemic species and the introduction of non-native species that already have a widespread abundance. Changes to the habitat may pro-mote both the extinction of native endemic species and the introduction of non-native species. The effects of habitat change will likely be similar in all urban environments as urban environments are all built to cater to the needs of humans.

The urban environment can also increase diversity in a number of ways. Many foreign organisms are introduced and dispersed naturally or artificially in urban areas. Artificial introductions may be intentional, where organisms have some form of human use, or accidental, where organisms attach themselves to transportation vehicles. Humans provide food sources (e.g. birdfeeder seeds, trash, garden compost) and reduce the numbers of large natural predators in urban environments, allowing large populations to be supported where food and predation would normally limit the population size. There are a variety of different habitats available within the urban environment as a result of differences in land use allowing for more species to be supported than by more uniform habitats.

Ways to Improve Urban Ecology: Civil Engineering and Sustainability

Cities should be planned and constructed in such a way that minimizes the urban effects on the surrounding environment (urban heat island, precipitation, etc.) as well as optimizing ecological activity. For example, increasing the albedo, or reflective power, of surfaces in urban areas, can minimize urban heat island, resulting in a lower magnitude of the urban heat island effect in urban areas. By minimizing these abnormal temperature trends and others, ecological activity would likely be improved in the urban setting.

Need for Remediation

Urbanization has indeed had a profound effect on the environment, on both local and global scales. Difficulties in actively constructing habitat corridor and returning biogeochemical cycles to normal raise the question as to whether such goals are feasible. However, some groups are working to return areas of land affected by the urban landscape to a more natural state. This includes using landscape architecture to model natural systems and restore rivers to pre-urban states.

Sustainability

Pipes carrying biogas produced by anaerobic digestion or fermentation of biodegradable materials as a form of carbon sequestration.

With the ever-increasing demands for resources necessitated by urbanization, recent campaigns to move toward sustainable energy and resource consumption, such as

LEED certification of buildings, Energy Star certified appliances, and zero emission vehicles, have gained momentum. Sustainability reflects techniques and consumption ensuring reasonably low resource use as a component of urban ecology. Techniques such as carbon recapture may also be used to sequester carbon compounds produced in urban centers rather continually emitting more of the greenhouse gas.

Urban Nature (Urban Open Space and Urban Greening)

On of the main methodes of improving the urban ecology is including in the cities more or less natural areas: Parks, Gardens, Lawns. These areas improve the health, the well being of the human, animal, and plant population of the cities. Generally they are called Urban open space (although this word not always mean green space), Green space, Urban greening.

A study published in Nature's Scientific Reports journal in 2019 found that people who spent at least two hours per week in nature, were 23 percent more likely to be satisfied with their life and were 59 percent more likely to be in good health than those who had zero exposure. The study used data from almost 20,000 people in the UK. The results remain the same, whether it was in one trip or multiple, and benefits increased for up to 300 minutes of exposure. The benefits applied to men and women of all ages, as well as across different ethnicities, socioeconomic status, and even those with long-term illnesses and disabilities.

People who did not get at least two hours — even if they surpassed an hour per week — did not get the benefits.

The study is the latest addition to a compelling body of evidence for the health benefits of nature. Many doctors already give nature prescriptions to their patients.

The study didn't count time spent in a person's own yard or garden as time in nature, but the majority of nature visits in the study took place within two miles from home. "Even visiting local urban green spaces seems to be a good thing," Dr. White said in a press release. "Two hours a week is hopefully a realistic target for many people, especially given that it can be spread over an entire week to get the benefit".

New Urbanism

New Urbanism is an urban design movement which promotes environmentally friendly habits by creating walkable neighborhoods containing a wide range of housing and job types. It arose in the United States in the early 1980s, and has gradually influenced many aspects of real estate development, urban planning, and municipal land-use strategies.

New Urbanism is strongly influenced by urban design practices that were prominent until the rise of the automobile prior to World War II; it encompasses ten basic principles such as traditional neighborhood design (TND) and transit-oriented development (TOD). These ideas can all be circled back to two concepts: building a sense of community and the development of ecological practices.

Market Street, Celebration.

The organizing body for New Urbanism is the Congress for the New Urbanism, founded in 1993. Its foundational text is the *Charter of the New Urbanism*, which begins:

> "We advocate the restructuring of public policy and development practices to support the following principles: neighborhoods should be diverse in use and population; communities should be designed for the pedestrian and transit as well as the car; cities and towns should be shaped by physically defined and universally accessible public spaces and community institutions; urban places should be framed by architecture and landscape design that celebrate local history, climate, ecology, and building practice".

New Urbanists support: Regional planning for open space; context-appropriate architecture and planning; adequate provision of infrastructure such as sporting facilities, libraries and community centres; and the balanced development of jobs and housing. They believe their strategies can reduce traffic congestion by encouraging the population to ride bikes, walk, or take the train. They also hope that this set up will increase the supply of affordable housing and rein in suburban sprawl. The *Charter of the New Urbanism* also covers issues such as historic preservation, safe streets, green building, and the re-development of brownfield land. The ten Principles of Intelligent Urbanism also phrase guidelines for new urbanist approaches.

Architecturally, new urbanist developments are often accompanied by New Classical, postmodern, or vernacular styles, although that is not always the case.

New Broad Street, Baldwin Park.

Until the mid 20th century, cities were generally organized into and developed around mixed-use walkable neighborhoods. For most of human history this meant a city that was entirely walkable, although with the development of mass transit the reach of the city extended outward along transit lines, allowing for the growth of new pedestrian communities such as streetcar suburbs. But with the advent of cheap automobiles and favorable government policies, attention began to shift away from cities and towards ways of growth more focused on the needs of the car. Specifically, after World War II urban planning largely centered around the use of municipal zoning ordinances to segregate residential from commercial and industrial development, and focused on the construction of low-density single-family detached houses as the preferred housing format for the growing middle class. The physical separation of where people live from where they work, shop and frequently spend their recreational time, together with low housing density, which often drastically reduced population density relative to historical norms, made automobiles indispensable for practical transportation and contributed to the emergence of a culture of automobile dependency.

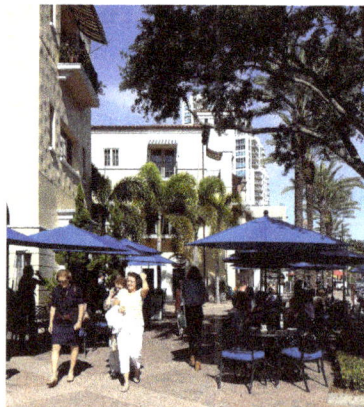

Beach Drive, St. Petersburg.

This new system of development, with its rigorous separation of uses, arose after World War II and became known as "conventional suburban development" or pejoratively as urban sprawl. The majority of U.S. citizens now live in suburban communities built in the last fifty years, and automobile use per capita has soared.

Celebration, FL Post Office.

Although New Urbanism as an organized movement would only arise later, a number of activists and thinkers soon began to criticize the modernist planning techniques being put into practice. Social philosopher and historian Lewis Mumford criticized the "anti-urban" development of post-war America. *The Death and Life of Great American Cities,* written by Jane Jacobs in the early 1960s, called for planners to reconsider the single-use housing projects, large car-dependent thoroughfares, and segregated commercial centers that had become the "norm". The French architect François Spoerry has developed in the 60's the concept of "soft architecture" that he applied to Port Grimaud, a new marina in south of France. The success of this project had a considerable influence and led to many new projects of soft architecture like Port Liberté in New Jersey or Le Plessis Robisson in France.

Rooted in these early dissenters, the ideas behind New Urbanism began to solidify in the 1970s and 80s with the urban visions and theoretical models for the reconstruction of the "European" city proposed by architect Leon Krier, and the pattern language theories of Christopher Alexander. The term "new urbanism" itself started being used in this context in the mid-1980s, but it wasn't until the early 1990s that it was commonly written as a proper noun capitalized.

In 1991, the Local Government Commission, a private nonprofit group in Sacramento, California, invited architects Peter Calthorpe, Michael Corbett, Andrés Duany, Elizabeth Moule, Elizabeth Plater-Zyberk, Stefanos Polyzoides, and Daniel Solomon to develop a set of community principles for land use planning. Named the *Ahwahnee Principles* (after Yosemite National Park's Ahwahnee Hotel), the commission presented the principles to about one hundred government officials in the fall of 1991, at its first Yosemite Conference for Local Elected Officials.

Calthorpe, Duany, Moule, Plater-Zyberk, Polyzoides, and Solomon founded the Chicago-based Congress for the New Urbanism in 1993. The CNU has grown to more than three thousand members, and is the leading international organization promoting New Urbanist design principles. It holds annual Congresses in various U.S. cities.

In 2009, co-founders Elizabeth Moule, Hank Dittmar, and Stefanos Polyzoides authored the Canons of Sustainable Architecture and Urbanism to clarify and detail the relationship between New Urbanism and sustainability. The Canons are "a set of operating principles for human settlement that reestablish the relationship between the art of building, the making of community, and the conservation of our natural world". They promote the use of passive heating and cooling solutions, the use of locally obtained materials, and in general, a "culture of permanence".

New Urbanism is a broad movement that spans a number of different disciplines and geographic scales. And while the conventional approach to growth remains dominant, New Urbanist principles have become increasingly influential in the fields of planning, architecture, and public policy.

Defining Elements

New Urbanism.

Andrés Duany and Elizabeth Plater-Zyberk, two of the founders of the Congress for the New Urbanism, observed mixed-use streetscapes with corner shops, front porches, and a diversity of well-crafted housing while living in one of the Victorian neighborhoods of New Haven, Connecticut. They and their colleagues observed patterns including the following:

A park in Celebration.

Great King St, New Town.

- The neighborhood has a discernible center. This is often a square or a green and sometimes a busy or memorable street corner. A transit stop would be located at this center.

- Most of the dwellings are within a five-minute walk of the center, an average of roughly 0.25 miles (0.40 km).

- There are a variety of dwelling types — usually houses, rowhouses, and apartments — so that younger and older people, singles and families, the poor and the wealthy may find places to live.

- At the edge of the neighborhood, there are shops and offices of sufficiently varied types to supply the weekly needs of a household.

- A small ancillary building or garage apartment is permitted within the backyard of each house. It may be used as a rental unit or place to work (for example, an office or craft workshop).

- An elementary school is close enough so that most children can walk from their home.

- There are small playgrounds accessible to every dwelling — not more than a tenth of a mile away.

- Streets within the neighborhood form a connected network, which disperses traffic by providing a variety of pedestrian and vehicular routes to any destination.

- The streets are relatively narrow and shaded by rows of trees. This slows traffic, creating an environment suitable for pedestrians and bicycles.

- Buildings in the neighborhood center are placed close to the street, creating a well-defined outdoor room.

- Parking lots and garage doors rarely front the street. Parking is relegated to the rear of buildings, usually accessed by alleys.

- Certain prominent sites at the termination of street vistas or in the neighborhood center are reserved for civic buildings. These provide sites for community meetings, education, and religious or cultural activities.

Anthropocene

Anthropocene Epoch is the unofficial interval of geologic time, making up the third worldwide division of the Quaternary Period (2.6 million years ago to the present), characterized as the time in which the collective activities of human beings (Homo sapiens) began to substantially alter Earth's surface, atmosphere, oceans, and systems of nutrient cycling. A growing group of scientists argue that the Anthropocene Epoch should follow the Holocene Epoch (11,700 years ago to the present) and begin in the year 1950.

Although American biologist Eugene Stoermer coined the term in the late 1980s, Dutch chemist and Nobelist Paul Crutzen is largely credited with bringing public attention to it at a conference in 2000 as well as in a newsletter printed the same year. In 2008 British geologist Jan Zalasiewicz and his colleagues put forth the first proposal to adopt the Anthropocene Epoch as a formal geological interval. In 2016 the Anthropocene Working Group of the International Union of Geologic Sciences (IUGS) voted to recommend the Anthropocene as a formal geologic epoch at the 35th International Geological Congress. In order for this interval to be made official, it first must be adopted by the IUGS and the International Commission on Stratigraphy.

Quaternary Period with the Anthropocene Epoch

Eonothem/ Eon	Erathem/ Era	System/ Period	Series/ Epoch	Stage/ Age	millions of years ago
Phanerozoic ↑ ↓	Cenozoic ↑ ↓	Quaternary ↑	Anthropocene[1]		
					1950 CE
			Holocene		0.0117
			Pleistocene	Upper	0.126
				Middle	0.781
				Calabrian	1.806
				Gelasian	2.588

The quaternary period, reconfigured to accommodate the anthropocene epoch.

The Scale of Human Activity

Changes in rock strata and the makeup of the fossils they contain are used to mark the boundaries between formal intervals of geologic time. Throughout Earth's history, periods of upheaval characterized by mass extinctions, changes in sea level and ocean chemistry, and relatively rapid changes in prevailing climate patterns are captured in the layers of rock. Often these periods mark the end of one interval and the beginning of another. The formalization of the Anthropocene hinges on whether the effects of humans on Earth are substantial enough to eventually appear in rock strata. Most scientists agree that the collective influence of humans was small before the dawn of the Industrial Revolution during the middle of the 18th century; however, advancements in technology occurring since then have made it possible for humans to undertake widespread, systematic changes that affect several facets of the Earth system.

Farms surrounding the town of Nørreby.

At present, human beings have a profound influence over Earth's surface, atmosphere, oceans, and biogeochemical nutrient cycling. By 2005, humans had converted nearly two-fifths of Earth's land area for agriculture. (Cultivated land accounted for one-tenth of the land surface, whereas roughly three-tenths were used for pasture). An additional one-tenth of Earth's land area was given over to urban areas by this time. According to some estimates, humans have harvested or controlled roughly one-quarter to one-third of the biomass produced by the world's terrestrial plants (net primary production) on a yearly basis since the 1990s. Such sweeping control over Earth's plant production has been attributed in large part to the development of a method of industrial nitrogen fixation called the Haber-Bosch process, which was created in the early 1900s by German chemist Fritz Haber and later refined by German chemist Carl Bosch. The Haber-Bosch process synthesizes ammonia from atmospheric nitrogen and hydrogen under high temperatures and pressures for use in artificial fertilizers and munitions. The industrialization of this process increased the amount of usable nitrogen in the world by 150 percent, which has greatly enhanced crop yields and, along with other technological developments, facilitated the exponential rise in the world's human population from about 1.6 billion–1.7 billion in 1900 to 7.4 billion by 2016.

As the human population grew, energy use increased, and energy derivation from wood and easily obtained fossil fuels (i.e., petroleum, natural gas, and coal) expanded. Carbon dioxide (CO_2) released by cooking fires and other sources during preindustrial times was dwarfed by the amount released by industrial furnaces, boilers, coal-fired power plants, gasoline-powered vehicles, and concrete production during the 20th and early 21st centuries. In the 1950s climate scientists began to track the annual increase in average global carbon dioxide concentrations in the atmosphere, which rose from approximately 316 parts per million by volume (ppmv) in 1959 to 390 ppmv a half century later. Many climatologists contend that the buildup of CO_2 in the atmosphere has contributed to a global rise in average surface temperatures of 0.74 °C (1.3 °F) between 1906 and 2005, loss of sea ice in the Arctic Ocean and the breakup of ice shelves along the Antarctic Peninsula, reduction in the size of mountain glaciers, changes in prevailing weather patterns, and more-frequent occurrence of extreme weather events in different parts of the world.

Furthermore, the oceans absorb much of the CO_2 released into the atmosphere by human activities, and this absorption has driven the process of ocean acidification. Seawater pH has fallen by 0.1 between about 1750 and 2010, a 30 percent increase in acidity. Marine scientists fear that continued increases in ocean acidity will slow, and possibly cease, the construction of reefs by corals in many parts of the world, dissolve the shells and skeletons of mollusks and corals, and interfere with the metabolic processes of larger marine animals. Since coral reefs are hubs of biodiversity in the oceans, the loss of coral will likely contribute to the demise of multitudes of other marine species either directly, through habitat loss, or indirectly, through changes in marine food chains. Other human-induced changes to the hydrosphere include the damming and diversion of rivers and streams, the rapid extraction of groundwater from freshwater aquifers, and the creation of large oxygen-depleted areas near the mouths of rivers.

Evidence in Layers of Rock

Many scientists who support the formalization of the Anthropocene Epoch argue that the effects of some of the changes mentioned above will create unique signatures in layers of rock. Some of these scientists argue that the increased rate of soil erosion from intensive agriculture and land-use conversion will leave a mark in rock strata, whereas others contend that such a mark will be barely noticeable and that other changes will be more apparent. For example, many scientists maintain that rising air temperatures at the surface brought about by global warming have caused glaciers and polar ice to melt and seawater to expand, both of which have contributed to a measurable rise in global sea level. Rising waters will change the stratigraphy in some places by submerging low-lying areas and allowing the ocean to deliver sediments farther inland than they do at present. Furthermore, as seawater pH declines, the depth at which carbonate minerals (e.g., limestone and chalk) form in the ocean will be shallower than it was during pre-industrial times. Many preexisting carbonate formations will dissolve in response to increases in ocean acidity, leaving a signature of striking dark layers of carbonate-depleted rock.

By far the most significant evidence of the Anthropocene in rock strata will be caused by a dramatic increase in extinctions occurring during this period. Several ecologists have noted that the rate of species extinction occurring since the middle of the 20th century has been more than 1,000 times that of the preindustrial period, comparable to the pace of other mass extinctions occurring over the course of Earth's history. The rapid extinction rate stems from the ongoing conversion of forests and other natural areas to agriculture and urban land and accelerated climate change resulting from alterations to the carbon cycle. As a result, it is expected that there will be stark differences in the fossils found in layers of rock deposited worldwide during preindustrial times and those that follow.

References

- Axelrod, yekaterina k.; diringer, michael n. (2008). "temperature management in acute neurologic disorders". Neurologic clinics. 26 (2): 585–603. Doi:10.1016/j.ncl.2008.02.005. Pmid 18514828

- Human-ecology, topic: britannica.com, Retrieved 16 January, 2019

- Laupland, kevin b. (2009). "fever in the critically ill medical patient". Critical care medicine. 37 (supplement): s273–s278. Doi:10.1097/ccm.0b013e3181aa6117. Pmid 19535958

- Krajick, kevin (2017-10-05). "ancient humans left africa to escape drying climate, says study". State of the planet. Columbia university. Retrieved 4 december 2018

- Singh, anita; agrawal, madhoolika (january 2008). "acid rain and its ecological consequences". Journal of environmental biology. 29 (1): 15–24. Pmid 18831326

- Wear, andrew (16 february 2016). "planning, funding and delivering social infrastructure in australia's outer suburban growth areas". Urban policy and research. 34 (3): 284–297. Doi:10.1080 /08111146.2015.1099523

- Anthropocene-epoch, science: britannica.com, Retrieved 29 March, 2019

Ecology of Animals 5

- **Animal Ecology**
- **Animal Diversity**

Animal ecology is concerned with the relationships of animals with their environments. It also studies the consequences of these relationships for evolution, population growth and regulation. The topics elaborated in this chapter will help in gaining a better perspective about these areas of animal ecology.

Animal Ecology

Animal ecology concerns the relationships of individuals to their environments, including physical factors and other organisms, and the consequences of these relationships for evolution, population growth and regulation, interactions between species, the composition of biological communities, and energy flow and nutrient cycling through the ecosystem. From the standpoint of population, the individual organism is the fundamental unit of ecology. Factors influencing the survival and reproductive success of individuals form the basis for under-standing population processes.

Two general principles guide the study of animal ecology. One is the balance of nature, which states that ecological systems are regulated in approximately steady states. When a population becomes large, ecological pressures on population size, including food shortage, predation and disease, tend to reduce the number of individuals. The second principle is that populations exist in dynamic relationship to their environments and that these relationships may cause ecological systems to vary dramatically over time and space. One of the challenges of animal ecology has been to reconcile these different viewpoints.

Populations depend on resources, including space, food, and opportunities to escape from predators. The amount of a resource potentially available to a population is generally thought of as being a property of the environment. As individuals consume resources they reduce the availability of these resources to others in the population. Thus, individuals are said to compete for resources. Larger populations result in a smaller share of resources per individual, which may lead to reduced survival and fecundity. Dense populations also attract predators and provide conditions for rapid transmission of contagious diseases, which generate pressure to reduce population size.

Changes in population size reflect both extrinsic variation in the environment that affects birth and death rates and intrinsic dynamics that result in oscillations or irregular fluctuations in population size. In some situations, the stable state may be a regular oscillation known as a limit cycle. Ecological systems also may switch between alternative stable states, as in the case of populations that are regulated at a high level by food limitation or at a low level by predators or other enemies. Switching between alternative stable states may be driven by changes in the environment.

Animal Population Ecology

Population is a group of organisms belonging to the same species occupying a particular space at a particular time. Individual members of a population are interbreeding and live in a particular place, in the same time and interact to one another as a society. A particular group of organisms, according to this definition, should fulfill the following points if it needs to be considered a population:

- Members must belong to the same kind of species.

- All the members must occupy the same place,

- Members must live in the same time,

- Members must interact to one another.

Biologists have for some time recognized that the most important level of organization of a species is its population, because at this level the gene pool is most coherent. Population is the basic unit of a community. Population ecology, hence, deals with the distribution and abundance of a population and the factors governing them.

Socializations in a Population

The degree of socialization or interaction between members of a population ranges from the most solitary animals like humming bird to the most complex interaction in social insects such as termites and honeybees. A society is a more complex interaction between members of a population. In a hive of honeybee or a mound of termites, millions of insects live in a very perfect coordination. They have complex division of labor and communication. Most of primates such as Chimpanzees and Gorillas are also highly social mammals. In those highly social animals (called Eusocials) rearing of youngs is in cooperative. The colony is also permanent by which offspring remain cooperative to the parents.

Importance of Socialization

Animals live in a society for their advantage. Some of the benefits include:

- Collective defense: Animals live in a society to protect from predators either by group fighting or confusing the predator. It is also useful to immunize the weak social groups such as female and young. In Buffalo, for example, members of the society protect predators using geometrical effect in that female and young are put in the middle, and all the other are standing in a circle pointing their face to the predator for physical defense. It is also important to spot the predator and communicate easily. Organizing in a society is useful for making confusion and also mobbing against tough predators.

- Improve foraging efficiency: Groups can forage more efficiently than solitary individuals by mutual vigilance for predators' defense. Each individual in a flock can get reduced vigilant time allowing more time for foraging. In an experiment, Ostriches raise their heads at frequent intervals 3-4 times per minute to scan the presence of predators in the area while they are feeding alone. However, the frequency is reduced to less than 2 times per minute per individual while they are feeding in a group of 3 or 4. Making the society helps them to spend their time for foraging than scanning their enemy.

- Social facilitation: Animals learn some of the behaviors by imitating from the action of the members of the flock. This helps to save time spent for learning through trial.

- Information transfer: Each member can communicate to each other about food sources and other matters. Example, the "waggle dance" in worker bees is a good example. The dance looks number "8", and the direction of the cross bar tells the direction of the food from the hive, and the length of the cross bar tells the distance from the hive.

Population Growth

Population is not static, but they are in change. The number of population may change from time to time. A universal characteristic of living things is that sexually mature individuals have the ability to produce ≥ 1 offspring. Thus, natural populations have the ability to grow. For example, each spring, in temperate oceans and lakes around the globe, planktonic populations of diatoms and algae take advantage of the increasing availability of sunlight and abundance of nutrients. Sizes of populations fluctuate in terrestrial, as well as aquatic, environments. The capacity for population growth, even among vertebrates, is enormous, a fact that is sometimes clearly demonstrated when species are introduced to new regions having suitable habitat. For example, when domestic sheep were introduced into Tasmania, a large island off the southeastern coast of Australia, the population increased from fewer than 200,000 in 1820 to more than 2 million by 1850.

Perhaps the most meaningful and dramatic example of the capacity of natural populations to grow is that of our own species. Although humans are not the most numerous animal species on Earth, the proliferation of the human species is, by any accounting, a remarkable ecological event. The human population began to expand rapidly after 1600, reaching 1 billion by the early 1800s, doubling that number by 1930 and doubling again by 1975. In 1995, the global human population reached 6 billion. There are currently 6.3 billion people alive on Earth today.

A population growth is determined largely by the difference between natality and mortality. Depending on the direction of movement, migration could also affect the population density of a given locality. There are two kinds of population growth models:

- Exponential growth models.

- Logistic growth models.

Exponential Population Growth Model

This model assumes that if there is no environmental constraint that hiders a population growth, as a result of which the population shows *geometric or exponential increase* until it overshoots the ability of the environment to support it. The growth curve is J-shaped. This kind of population growth is common in new habitats where here is no shortage of resources. In this kind of growth, there is exponential relationship between time "t" and the population number at time "t". This is characteristic of r-selected species adapted to new (unstable) and resource-rich environment. A good example is binary fission in bacteria. A bacteria cell reproduces asexually by binary fission in which two identical daughter cells are reproduced from a parent cell by mitotic division. If there is good nutrient, optimum pH and favorable environment for growth, the daughter cells continue to grow and divide the same way every 20 minutes.

The following table shows the population increase of bacteria through binary fission. Suppose the initial population size is 2 and if we assume no death and migration of individuals, after 1 generation each cell divides in to two and the population size becomes 4. At the second generation, the population size grows to 8. At the 3rd generation, the size grows to 16, and 32, and so on.

Table: The arithmetic increase of a bacterial population through binary fission.

Number of Generations / time- (t)	0	1	2	3	4	5	6	7	8	9	10
Total population size (N_t)	2	4	8	16	32	64	128	256	1012	2024	4048
Relationship between N_t and "t"	$2(2^0)$	$2(2^1)$	$2(2^2)$	$2(2^3)$	$2(2^4)$	$2(2^5)$	$2(2^6)$	$2(2^7)$	$2(2^8)$	$2(2^9)$	$2(2^{10})$

From this data, if we draw a graph by putting "t" on the x-axis and N_t on the y-axis, we can see a clear "J" shaped curve as shown in figure, expressing the exponential relationship to one another.

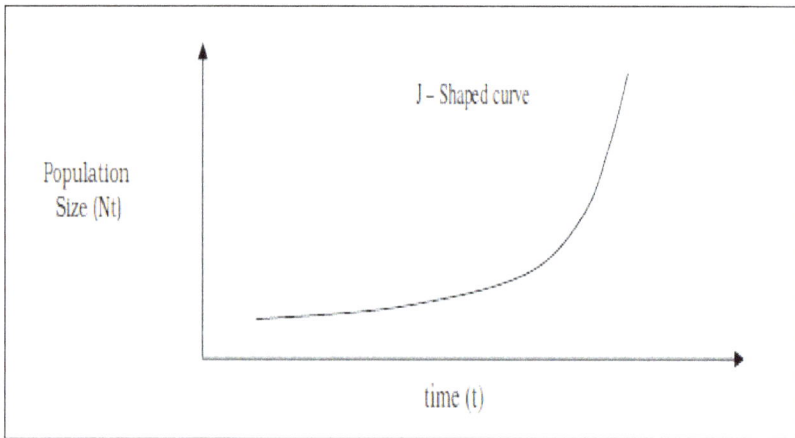

The pattern of population growth according to the exponential growth model.

The relationship between the population and the number of generations (or time) could be summarized by the following formula:

$$N_t = N_0 2^t$$

Where, N_t is the total number of population after time "t", N_0 is the starting population at time 0, and "t" is the number of generation (time of reproduction).

In the above example, however, we did not consider the effect of death and migration on the population size. If we consider all the natural processes, the net natural increase, also called intrinsic rate of growth or biotic potential ("r"), would be the difference between Birth rate (N) and Death rate (M), and also rate of net migration (Immigration (I) – Emigration (E)), as given by:

$$r = (N-M) + (I - E) \text{ or } (N+I) - (M+E)$$

The overall population increase using exponential model is, therefore, given by the following equation:

$$N_t = N_0 e^{rt}$$

Where "e" is the base for natural logarithm given by a constant 2.71828; "r" is the intrinsic population increase, and "t" is time or number of generations.

Example: If the existing Ethiopian livestock population (N_0) is 80 million and the birth rate (N) and death rates (M) are 45 and 15 individuals out of 1,000 populations, respectively, calculate the total population size (N_t) after 10 years.

$$N_{10} = 80000,000 \ (e^{\ (0.045-0.015)10})$$

$$= 8(10^7) \times 2.71828^{(0.3)}$$

= 80,000,000 (1.34985)

= 107,988,682.8

Logistic Growth Model

This growth model assumes that most populations cannot continue to grow exponentially because of the resistance coming from their environment that prevents from further growth. As the result, the population grows exponentially until it reaches the carrying capacity (K). However, as the population approaches to the carrying capacity, the growth is limited by the resistant factor hence the population increases arithmetically. Carrying capacity is the maximum number of individuals that a habitat could support. This kind of growth is the characteristic of k-selected species adapted to stable environment, where there is competition for resources.

Populations invading a new area where space and food are plenty will undergo exponential growth to begin with. But, due to different external constraints coming from the environment like shortage of space, food and other stresses, the rate of increase start to decline. The curve in the graph is called "S" or "Sigmoid" curve as shown in the diagram on figure. The rate of population increase at any time is density dependent.

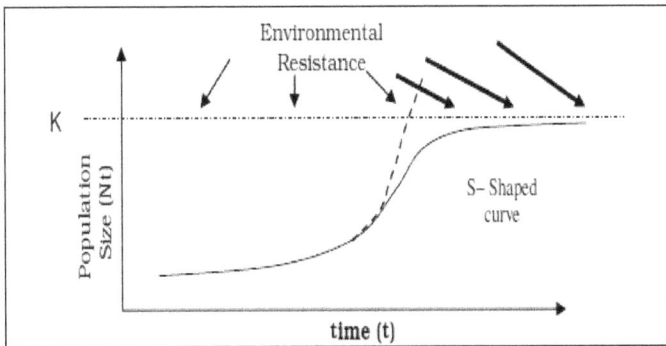

The pattern of population growth according to the logistic growth model.

Suppose an environment has a carrying capacity given by K for a particular population, the intrinsic rate of natural increase "r" is progressively reduced as population size approaches towards "K"

The available niche in the habitat, which is the inverse measure of environmental "resistance" or the "effect of crowding", is given by:

$$\frac{K - N}{N}$$

Where, "N" is the population at time "t", and "K" is the carrying capacity of the habitat. When "N" is smaller, the resistance value is nearly zero. That is there are numerous opportunities for the population. Hence, the niche is empty without any resistance that the biotic potential (r) is fully realized. In this case, the population can grow exponentially. However, when N is higher and closer to the carrying capacity (K), the value approaches to o; there is no available resource in the habitat to accommodate a single individual. This shows that at any ecosystem the population growth is limited by the degree of the environmental resistance. The integrated equation of logistic growth is, therefore, given by:

$$Nt = \frac{K}{1 + \left[\dfrac{K - N_0}{N_0} \right] \times e^{-rt}}$$

Population Characteristics

A population, as a group, has unique characteristics, which can be statistical measured, and cannot be applied to individual organisms. The three basic group characteristics of a population are:

- Density,
- Primary population parameters,
- Secondary population parameters.

Density

It is defined as the number of individuals of a population per a given unit area. For example, 5,000 individual of cattle per a square kilo meter of range land; 300 lions per square kilometer of rangeland; or 500 browse plants per hectare of forestland. It could be also expressed in terms of a unit volume (e.g., 10 million bacteria per a cubic centimeter of water).

Density varies with time and space. Individuals in natural populations are affected by their density in some way. For example, if the population density is high, resources are shared unequally. On the other hand, if the density is very low, it reduces the chance of performing behavioral activities essential to the welfare of the population. For example, trees in crowded stands grow more slowly due to shortage of water, nutrients and sufficient light. The same way in animals, very high population density affects the population in various way such as by causing scarcity on available food resources and the access to nest site (e.g., in birds), and also aggravating the spread of disease. Having too few individuals in a population may also affect the population by reducing their chance of finding a mate.

Density Dependent (Primary) Population Parameters

This category includes those characteristics of a population that are density dependent, i.e, their magnitude directly affect the density of a certain population. This includes:-

- Birth rate (natality),
- Death rates (mortality),

- Immigration,

- Emigration.

Birth Rate

This is the number of new individuals added to the population by birth per a given number of populations (usually 1,000 or 10,000) per year. New individuals could be added through,

- Birth,

- Hatching,

- Germination,

- Fission.

A certain population has its own unique birth rate that differentiates itself from the other. For example, the birthrate in Ethiopian cattle population is different from that of Sweden. This is largely affected by the rate of fertility and fecundity of the population.

Fertility indicates the physiological capability of organisms for breeding, while fecundity is an ecological concept that expresses the number of offerings produced during a period of time by the population.

Death Rate

It refers to the number of individuals removed out of a population through death. It decreases the density of a population. It is affected by a number of environmental factors such as,

- Competition,

- predation,

- disease,

- Hereditary characteristics, etc.

The specific age of an organism it lives in a population before its death is referred to Longevity. There are two types of longevity:

- Potential longevity: It is the maximum life span attainable by an individual of a particular species. The limit is set by the physiology of the organism.

- Realized longevity: It is the average longevity of the individuals in a population living under real environmental conditions. For example, the European robin has an average life expectancy of 1 year in the wild and 11 years in captivity.

Dispersal

Dispersal is the movement of animals outside their normal home range for different reasons such as:

- Reproduction,

- Survival,

- Spread.

The dispersal could be temporary or permanent in a population. Dispersal is important to prevent inbreeding and it facilitates gene flow between local populations. There are two ways of migration in a population.

- Emigration: It is dispersal or migration of individuals away from the population. Individuals leaving a population are called emigrants.

- Immigration: It is dispersal or migration of individuals into a population. Individuals joining a population coming from another group are called immigrants. It increases the population size.

Dispersal is common in mobile animals such as mammals, birds, fish, and insects. In the case of plants, the main dispersal units are seeds or spores. Most seeds and spores do not travel very far. However, some seeds can travel very long distance with the help of different vectors.

Sex Ratio

It is the ratio of male to female individuals in a population. The natural tendency of male to female sex ratio is 1 to1 (1:1). However this may be affected by different factors and hence affects the population size. Sex ratio changes with age due to differences in the mortality rate of male and female individuals. For example, the death rate is higher for males than female in human population, as the life expectancy of male is lower than the female.

Mating System

The type of matting between different mating types (sexes) in a population affects the population density.

Density Independent (Secondary) Population Parameters

This category includes those population characteristics that are not associated to the density of a given population. These factors do not necessarily affect the density of a given population, hence called density independent characteristics. These include:

- Age distribution,

- Genetic composition,

- Spatial distribution of individuals.

Dispersion Pattern

Dispersion is the spatial pattern of individuals in a population relative to one another. There are tree basic patterns of dispersion: Regular, Random and Clumped.

Random distribution- In this kind of dispersion, the position of one individual in a population is unrelated to the positions of its neighbors. Individuals of a population are distributed randomly. Even if this kind of dispersion is rare in nature it is evident in forest trees, and invertebrates of forest floor.

- Regular/Even/distribution – Individuals of a population are more or less equidistant from one another. It is rare in nature but common in managed systems like in crop plantation or territorial animals. This may occur if there is inter-specific competition arising from scarcity of resources such as, if there is competition for crown or root space among forest trees or for moisture among desert plants.

- Clumped/Aggregate/distribution- In this kind of dispersion, individuals of a population are aggregated into patches. This could arise from a) socialization tendency of individuals, b) clumber nature of resource distribution in nature or c) the tendency of offspring to remain into close distance with their parents.

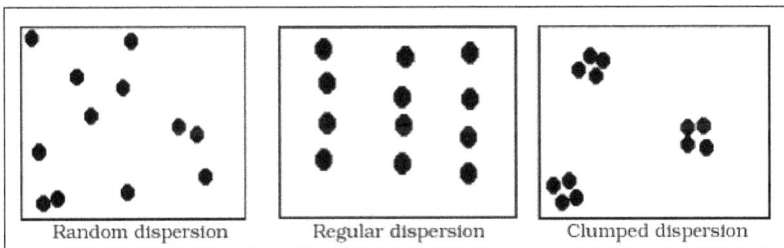

| Random dispersion | Regular dispersion | Clumped dispersion |

Common types of spatial distribution of individuals in a population.

Age Structure

This refers to the proportion of individuals in each age group. It represents the ratio of the various age classes in a population to each other at a given time. Based on the potential of individuals to reproduce, we classify individuals of a population into three groups: pre-reproductive, reproductive and post-reproductive. Under a normal condition, the age structure of most populations has a pyramid shape, which is broader at the base and narrower at the tip. If the proportion of the post reproductive age group is higher than the other groups, for example, this population is declining. However, if it is highly broad at the base, the population has a great momentum to grow fast, as the young individuals enter the reproductive age group.

Age Determination in Animals

There are different ways to determine the age of individuals of animal populations. The common ways include the following:

- The wear and replacement of teeth in ungulates.

- The growth rings in the cementum of teeth of carnivores and ungulates.

- The annual growth rings in the horns of artiodactyls.

- The weight of eye lens in rabbits which increase with age.

- Plumage changes and wears in birds, etc.

Age Determination in Plants

There are different ways of determining the age of a plant. The common ways include the following:

- Diameter at breast height - It is estimation based on the assumption that stem diameter of plants increase with age. The greater the diameter the older the tree. It applied to the higher trees.

- Counting annual growth rings - The number of rings is used to determine the number of dormant seasons, hence the age. However, this is applicable only for temperate trees. Because of winter dormancy followed by spring burst of growth, temperate plants show distinct growth pattern in the form of rings that corresponds with flowering or leaf emergency.

Population Limitation and Regulation

Populations are often regulated by environmental factors. Population increase depends on the reproductive fitness and lifespan of individual members. Factors such as:-

- Nutrient availability,

- Flood,

- Drought,

- Predators,

- Disease affects the population size.

These environmental factors are classified into three categories depending on:

- The degree of their variability in time,

- The degree of predictability of the variation.

Constantly Limiting Factors

This includes those factors that are always in short supply and are relatively constant. These factors have fairly constant magnitude, as a result of which individuals of a population have to compete for those scares resources. These kinds of factors therefore, always limit the size of the population. Examples include light, moisture and space for plants; and nesting site and food for birds. These kinds of factors do not usually produce large changes in a population.

Variably Limiting Factors

These factors are variable but predictable. This includes seasonal drought or cold, or variation in food availability. These factors influence the population only at a certain times of the year or a certain period, and cause the population to crash. Some organisms have developed a way of escaping this problem. Migration of some birds and mammals, and dropping of leaves in deciduous trees in winter time are some of the examples.

Unpredictable Factors

This includes those factors that do not have regularity and predictability. Hence it changes the population considerably over time. Weather effects, grazing or predation pressure, disease, fire, and volcanic eruption are some of the examples. These factors may kill most of the population, all of a sudden.

Animal Diversity

Kingdom Animalia is composed of a range of organisms united by a set of common characteristics. Barring a few exceptions, animals are multicellular eukaryotes that move, consume organic matter, and reproduce sexually. Although these attributes are shared, species within this kingdom are also extremely diverse. This diversity is due to adaptation of each species to a different niche. The niche of a species includes the area, function, and interrelationship of that species with other biotic and abiotic factors in its environment. Niche specialization through evolutionary adaptation allows species to survive and reproduce effectively in their environment and reduces competition among species within the same habitat.

Classification

Diversity in Kingdom Animalia has led to the classification of 36 distinct phyla, based on their evolutionary lineage. Seven of these phyla will be discussed, including Porifera, Cnidaria, Platyhelminthes, Annelida, Mollusca, Arthropoda, and Chordata. The first, phylum Porifera, is the earliest, simplest, most ancestral phylum. It includes multicellular, asymmetrical filter feeding sponges that lack distinct tissue layers. Interestingly, sponges can regrow after being broken apart to the level of a single cell. Next, increasing in complexity, the phylum Cnidaria includes jellyfish and coral. Members of this phylum are radially symmetric and diploblastic (possessing two tissue layers). These tissue layers are called the endoderm, which makes up the inner layer, and the ectoderm, which forms the outer layer. Organisms in phylum Cnidaria contain a simple gut with a single opening through which food enters and waste is excreted. Both Porifera and Cnidaria include organisms that are largely sedentary as adults. However, the remaining phyla have evolved adaptations that allow movement and more sophisticated behavior.

Following Cnidaria, there are two phyla of worms, termed Platyhelminthes and Annelida. Worms from both of these phyla breathe by exchanging gases across their skin. As a result, most worms have a large surface area to volume ratio and live in moist or aquatic environments. The skin of these organisms must remain moist to facilitate effective diffusion of gases. Another common characteristic of worms is their bilateral symmetry. This means that an organism can be symmetrically divided in half along a plane. Bilateral symmetry is seen not only in worms, but also in every phyla from this evolutionary point forward. Such symmetry is important to allow directed movement and to concentrate sensory organs near the anterior tip, or head, where they interact with their surroundings. Worms and later phyla also contain an additional tissue layer called the mesoderm that lies between the ectoderm and endoderm. As a result, these organisms are considered "triploblastic".

Of the two phyla of worms, Platyhelminthes is the more ancient and simple. These include flatworms and trematodes, which may be free-living or parasitic. Like Cnidarians, they possess a simple gut with a single opening. Phylum Annelida is more complex, including earthworms, leaches, and polychaeta. These organisms contain a body cavity, called the coelom, between the gut and body wall. They are also distinct due to the segmentation of their bodies. Division of an organism into segments was an important evolutionary step, allowing adaptation of each segment to specialized forms and functions. Major evolutionary advances in this phyla include the development of anterior segments with appendages specialized for grasping and processing food as well as posterior segments with appendages specialized for locomotion.

The next phylum, Mollusca, contains highly morphologically diverse organisms, including octopuses, snails, slugs, and clams. There are few defining features for this group. However, all mollusks possess a mantle, which acts as a protective cover for the respiratory organs, digestive tract, and reproductive structures. Next, the most diverse group of organisms is phylum Arthropoda. This phyla alone makes up approximately three quarters of all living and extinct species known today, including crustaceans, insects, and arachnids. All members of this phylum have a hard, outer coating made of chitin, called an exoskeleton. This exoskeleton is periodically shed for a new, larger one to take its place. Like Annelida, arthropods are segmented. However, instead of multiple repeating sections, arthropods are broken up into three distinct segments called the head, thorax, and abdomen. Most have specialized sensing structures attached to the head called antennae.

Lastly, phylum Chordata includes all animals with a backbone and a few without. Two defining features that unite this group are the presence of a nerve chord and a notochord. The nerve cord is a major part of the chordate nervous system and the notochord is a cartilaginous structure that runs between the nerve cord and the digestive tract. Phylum Chordata includes tunicates (which lose their nerve cord and notochord after metamorphosis from the larval to adult life stage), fish, amphibians, birds, reptiles, and mammals (including humans).

Form and Function

The vast diversity in body structure, habitats, and behavior observed across all 36 phyla within Kingdom Animalia is truly astounding. To understand these differences better, it is important to consider the connection between form and function. "Form" simply means the shape, size, and substance of a structure or organism while "function" stands for the way in which that structure is used. Accordingly, the function of a structure can be predicted based on the form observed, or vice versa. For example, a bird specialized to crack hard nuts would be predicted to have a sturdy, sharp beak while a bird that eats fruit may have a smaller, more rounded beak.

Importantly, the form of an organism is constrained by its physiological requirements. For instance, all animals must consume and digest organic matter and exchange gas with their environment. As a result, any adaptations to an animal's form must continue to facilitate these processes efficiently. For example, in smaller animals, diffusion of gases across the skin is possible and often useful. However, as the size of an animal increases, its surface area to volume ratio decreases until diffusion across the skin can no longer provide sufficient gas exchange. Such larger animals have developed complex systems to deal with gas exchange, such as the lungs of a whale or the gills of a fish. As a result, the morphology of gas exchange systems depends on both the size of an organism and its environment.

Similarly, it generally follows with other types of adaptations that form follows function, constrained only by the physiological requirement of the organism. However, there are cases in which form does not follow function. Vestigial structures are those that served a purpose in an organism's ancestor but are no longer useful in their current form. Such structures have little to no impact on the fitness of an individual, and are thus not evolutionarily selected for or against within the population. As a result, some structures persist with limited functional capacity. For example, the human appendix, a small pocket at the end of the large intestine, is an evolutionary remnant from our ancestors. While this structure has been hypothesized to store beneficial gut bacteria, removal of the appendix is possible with no deleterious effects.

Convergent Evolution

Among adaptations seen in Kingdom Animalia, many have developed multiple times throughout evolutionary history. Consider the wings of a bat and those of a bird as an example. These structures are not a result of evolutionary relatedness, but have instead arisen from convergent evolution[4]. Convergent evolution occurs when two organisms independently evolve similar traits due to similar environmental pressures or niches. The wings of bats, birds, and insets have all developed independently and are said to be "convergent traits". Clearly, such traits are quite beneficial for the niches in which they arise.

Aquatic Ecology 6

- **Freshwater Ecosystem**
- **Lake Ecosystem**
- **River Ecosystem**
- **Marine Ecosystem**

Aquatic ecology refers to the study of plants and animals living in water and their surroundings. Aquatic ecosystem, freshwater ecosystem, lake ecosystem, river ecosystem and marine ecosystem are a few branches of aquatic ecology. This chapter has been carefully written to provide an easy understanding of these branches of aquatic ecology.

Aquatic ecology includes the study of relationships between individuals of the same species, between different species, and between organisms and their physical and chemical environments in all aquatic environments, including oceans, estuaries, lakes, ponds, wetlands, rivers, and streams.

Communities of plants and animals living in water are known as aquatic ecosystems. They are divided into two main groups. Freshwater ecosystems are found in water containing low concentrations of salts, from ponds to estuaries. Marine ecosystems are found in the saltwater of seas and oceans. Most of us are not far away from an aquatic ecosystem of some kind, whether it be in the ocean or a local pond.

The nature of an aquatic ecosystem is shaped, as on land, by the availability of food, oxygen, and the prevailing temperature. Added to this is salinity, which is the salt concentration of the water. Aquatic ecosystems in shallow waters, where there is plenty of sunlight, generally tend to be the most productive. Water pollution, generally coming from human activities, comprises the greatest pressure on aquatic ecosystems. For instance, fish can be killed by acid rain in lakes or lack of oxygen where excess nutrients have been dumped in an estuary.

Aquatic ecosystems are the communities of organisms, their surrounding watery environment, and the relationship between the two. Aquatic, or watery, environments are divided into freshwater and marine. Freshwater has less than one gram per liter of dissolved solids, mainly salts, of which sodium chloride is the most important as far as living organisms are concerned. Salinity is the term given to the concentration of salts in a body of water. Saltwater contains more than one gram per liter of dissolved solids.

Freshwater ecosystems are found in ponds, lakes, reservoirs, rivers, and streams. Marine, or saltwater, ecosystems are found in the seas and oceans. Estuaries, which are places where a river meets the sea, such as San Francisco Bay, are part freshwater and part marine in their makeup. The diversity of an aquatic ecosystem depends upon temperature, availability of light, nutrients, oxygen, and salinity. This makes for very different ecosystems in, for instance, a shallow pond in a temperate climate with plenty of light compared to the deep ocean where conditions are dark and cold.

A wide range of plants, animals, and microbes are found in aquatic ecosystems, from bacteria that tolerate high temperatures in hydrothermal vents at the bottom of the ocean to the blue whale, which is the world's largest animal. The smallest are the phytoplankton and zooplankton, the microscopic animals and plants forming the bottom layer of the aquatic food chain. There are also many aquatic invertebrates including worms, insects, and crustaceans. Among the aquatic vertebrates, amphibians, such as frogs, live on land and water, while fish are purely water-dwelling animals. Many birds, such as kingfishers, gulls, and ducks, live on or near water. Whales are one of the few mammals that can live in the marine environment.

Rivers and streams are lotic, or flowing, freshwater environments. Their organisms tend to be small, with flattened bodies, so they are not swept away. Fallen leaves, insects, and other detritus are important food sources. Lakes and reservoirs are lentic, or layered, systerms with still water. The littoral layer, near the edge of a lake, generally has a large ecosystem with organisms that can use both land and water such as dragonflies, frogs, ducks, and turtles.

Estuaries are regions where the river meets the sea, or ocean. They comprise a semi-enclosed coastal area, open to the marine environment. An estuary has different ecosystems, such as a salt marsh, which is farthest from the sea. Mudflats are tidal and are therefore, sometimes aquatic and sometimes terrestrial. The range of species in mudflats tends to be narrow, but there can be many individuals within this range. The channels are the areas of an estuary that connect to the ocean and are generally rich in fish and crustaceans like crabs. Overall, an estuary is a rich ecosystem because it tends to be shallow and there is a good mixing of nutrients.

The main differences between freshwater and marine environments are the salinity of the water, the depth of the water, and the availability of sunlight for photosynthesis. Freshwater organisms need mechanisms to prevent water loss, because water flows from regions of high salt concentration to those of low salt concentration. Marine organisms at depth have to deal with the pressure of water on their bodies and lack of sunlight. These factors mean that freshwater and marine ecosystems can be very different in their diversity.

Thousands of different invertebrates are found in the seas and oceans. There are sponges, echinoderms (which include starfish, sea urchins, and sea cucumbers), and cnidarians, such as jellyfish, corals, and sea anemones. Mollusks, which are found on land and sea, include the gastropods (limpets, slugs, and snails), cephalopods (octopus, squid, and cuttlefish), and bivalves (clams, mussels, cockles, oysters, and scallops). Most crustaceans are marine animals and include crabs, lobsters, shrimps, barnacles, and woodlice. There are also many species of marine worms living on the sea floor or in the sand of coastal areas.

Around half of the 25,000 known species of fish live in the marine environment, mainly in shallower, warmer.

Ecosystem: The community of individuals and the physical components of the environment in a certain area.

Euphotic zone: The uppermost layer of a body of water in which the level of sunlight is sufficient for photosynthesis to occur.

Eutrophication: The process whereby a body of water becomes rich in dissolved nutrients through natural or man-made processes. This often results in a deficiency of dissolved oxygen, producing an environment that favors plant over animal life.

Freshwater: Water containing less than one gram per liter of dissolved solids.

Littoral: The region of a lake near the shore.

Lotic: Flowing water, as in rivers and streams.

Symbiosis: A pattern in which two or more organisms of different species live in close connection with one another.

Meanwhile, whales are marine mammals, breathing with lungs, and the largest animals in the oceans. In the open ocean, the nature of an ecosystem is dependent on depth. In the euphotic zone, there is net primary production of food by photosynthesis carried out by 4,000 or so species of phytoplankton. These are eaten by slightly larger animals, such as tiny crustaceans called copepods. These, in turn, will be eaten by animals bigger than they are. Farther down, at 4 miles (6.5 km) deep and beyond, there is no light in these zones close to the ocean floor except for that which is emitted from the organisms themselves. Thousands of species, including bacteria, squid, and fish, emit flashes of light by a process known as bioluminescence. They feed on one another and on detritus falling from the upper layers.

The marine environment contains some important ecological niches. For instance, coral reefs consist of animals called stony corals, each of which is a polyp with tentacles that can trap organisms. These polyps live in symbiosis with photosynthetic algae. Coral reefs tend to occur in warm, shallow, clear waters where they provide a home for a diverse community of fish, worms, and crustaceans.

Impacts and Issues

Many human activities threaten the health of aquatic ecosystems. For instance, acid rain created from sulfur and nitrogen oxide emissions have turned many lakes and streams acidic, so they no longer support various fish species. Meanwhile, the building of dams to create hydroelectric power plants can block the routes of migratory fish like salmon.

In the marine environment, the coral reefs are among the world's most threatened ecosystems. They are affected by a range of factors, including destructive fishing practices, pollution, sewage, and global warming. Estuaries and shore areas are also at risk from pollution, which can cause eutrophication by raising nutrient levels in the water. Eutrophication encourages the growth of plant decomposers, which consume available oxygen in the water, affecting fish, and other marine organisms.

Freshwater Ecosystem

Freshwater ecosystems are a subset of Earth's aquatic ecosystems. They include lakes and ponds, rivers, streams, springs, bogs, and wetlands. They can be contrasted with marine ecosystems, which have a larger salt content. Freshwater habitats can be classified by different factors, including temperature, light penetration, nutrients, and vegetation.

Freshwater ecosystems can be divided into lentic ecosystems (still water) and lotic ecosystems (flowing water).

Limnology (and its branch freshwater biology) is a study about freshwater ecosystems. It is a part of hydrobiology.

Original attempts to understand and monitor freshwater ecosystems were spurred on by threats to human health (ex. Cholera outbreaks due to sewage contamination). Early monitoring focused on chemical indicators, then bacteria, and finally algae, fungi and protozoa. A new type of monitoring involves quantifying differing groups of organisms (macroinvertebrates, macrophytes and fish) and measuring the stream conditions associated with them.

Threats to Freshwater Ecosystems

Five broad threats to freshwater biodiversity include overexploitation, water pollution, flow modification, destruction or degradation of habitat, and invasion by exotic species. Recent extinction trends can be attributed largely to sedimentation, stream fragmentation, chemical and organic pollutants, dams, and invasive species. Common chemical stresses on freshwater ecosystem health include acidification, eutrophication and copper and pesticide contamination.

Extinction of Freshwater Fauna

Over 123 freshwater fauna species have gone extinct in North America since 1900. Of North American freshwater species, an estimated 48.5% of mussels, 22.8% of gastropods, 32.7% of crayfishes, 25.9% of amphibians, and 21.2% of fish are either endangered or threatened. Extinction rates of many species may increase severely into the next century because of invasive species, loss of keystone species, and species which are already functionally extinct (e.g., species which are not reproducing). Even using conservative estimates, freshwater fish extinction rates in North America are 877 times higher than background extinction rates (1 in 3,000,000 years). Projected extinction rates for freshwater animals are around five times greater than for land animals, and are comparable to the rates for rainforest communities.

Biomonitoring

Current freshwater biomonitoring techniques focus primarily on community structure, but some programs measure functional indicators like biochemical (or biological) oxygen demand, sediment oxygen demand, and dissolved oxygen. Macroinvertebrate community structure is commonly monitored because of the diverse taxonomy, ease of collection, sensitivity to a range of stressors, and overall value to the ecosystem. Additionally, algal community structure (often using diatoms) is measured in biomonitoring programs. Algae are also taxonomically diverse, easily collected, sensitive to a range of stressors, and overall valuable to the ecosystem. Algae grow very quickly and communities may represent fast changes in environmental conditions.

In addition to community structure, responses to freshwater stressors are investigated by experimental studies that measure organism behavioural changes, altered rates of growth, reproduction or mortality. Experimental results on single species under controlled conditions may not always reflect natural conditions and multi-species communities.

The use of reference sites is common when defining the idealized "health" of a freshwater ecosystem. Reference sites can be selected spatially by choosing sites with minimal impacts from human disturbance and influence. However, reference conditions may also be established temporally by using preserved indicators such as diatom valves, macrophyte pollen, insect chitin and fish scales can be used to determine conditions prior to large scale human disturbance. These temporal reference conditions are often easier to reconstruct in standing water than moving water because stable sediments can better preserve biological indicator materials.

Lake Ecosystem

A lake ecosystem includes biotic (living) plants, animals and micro-organisms, as well as abiotic (nonliving) physical and chemical interactions.

Lake ecosystems are a prime example of lentic ecosystems. Lentic refers to stationary or relatively still water, from lentus, which means sluggish. Lentic waters range from ponds to lakes to wetlands. Lentic ecosystems can be compared with lotic ecosystems, which involve flowing terrestrial waters such as rivers and streams. Together, these two fields form the more general study area of freshwater or aquatic ecology.

Lentic systems are diverse, ranging from a small, temporary rainwater pool a few inches deep to Lake Baikal, which has a maximum depth of 1642 m. The general distinction between pools/ponds and lakes is vague, but Brown states that ponds and pools have their entire bottom surfaces exposed to light, while lakes do not. In addition, some lakes become seasonally stratified. Ponds and pools have two regions: the pelagic open water zone, and the benthic zone, which comprises the bottom and shore regions. Since lakes have deep bottom regions not exposed to light, these systems have an additional zone, the profundal. These three areas can have very different abiotic conditions and, hence, host species that are specifically adapted to live there.

Important Abiotic Factors

Light

Light provides the solar energy required to drive the process of photosynthesis, the major energy source of lentic systems. The amount of light received depends upon a combination of several factors. Small ponds may experience shading by surrounding trees, while cloud cover may affect light availability in all systems, regardless of size. Seasonal and diurnal considerations also play a role in light availability because the shallower the angle at which light strikes water, the more light is lost by reflection. This is known as Beer's law. Once light has penetrated the surface, it may also be scattered by particles suspended in the water column. This scattering decreases the total amount of light as depth increases. Lakes are divided into photic and aphotic regions, the prior receiving sunlight and latter being below the depths of light penetration, making it void of photosynthetic capacity. In relation to lake zonation, the pelagic and benthic zones are considered to lie within the photic region, while the profundal zone is in the aphotic region.

Temperature

Temperature is an important abiotic factor in lentic ecosystems because most of the biota are poikilothermic, where internal body temperatures are defined by the surrounding system. Water can be heated or cooled through radiation at the surface and conduction to or from the air and surrounding substrate. Shallow ponds often have a continuous temperature gradient from warmer waters at the surface to cooler waters at the bottom. In addition, temperature fluctuations can vary greatly in these systems, both diurnally and seasonally.

Temperature regimes are very different in large lakes. In temperate regions, for example, as air temperatures increase, the icy layer formed on the surface of the lake breaks up, leaving the water at approximately 4 °C. This is the temperature at which water has the highest density. As the season progresses, the warmer air temperatures heat the surface waters, making them less dense. The deeper waters remain cool and dense due to reduced light penetration. As the summer begins, two distinct layers become established, with such a large temperature difference between them that they remain stratified. The lowest zone in the lake is the coldest and is called the hypolimnion. The upper warm zone is called the epilimnion. Between these zones is a band of rapid temperature change called the thermocline. During the colder fall season, heat is lost at the surface and the epilimnion cools. When the temperatures of the two zones are close enough, the waters begin to mix again to create a uniform temperature, an event termed lake turnover. In the winter, inverse stratification occurs as water near the surface cools freezes, while warmer, but denser water remains near the bottom. A thermocline is established, and the cycle repeats.

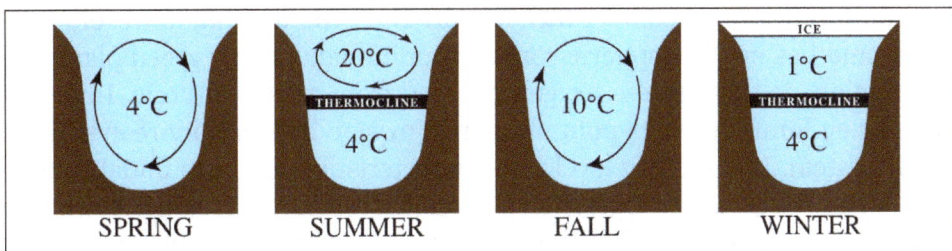

Seasonal stratification in temperate lakes.

Wind

In exposed systems, wind can create turbulent, spiral-formed surface currents called Langmuir circulations. Exactly how these currents become established is still not well understood, but it is evident that it involves some interaction between horizontal surface currents and surface gravity waves. The visible result of these rotations, which can be seen in any lake, are the surface foamlines that run parallel to the wind direction. Positively buoyant particles and small organisms concentrate in the foamline at the surface and negatively buoyant objects are found in the upwelling current between the two rotations. Objects with neutral buoyancy tend to be evenly distributed in the water column. This turbulence circulates nutrients in the water column, making it crucial for many pelagic species, however its effect on benthic and profundal organisms is minimal to non-existent, respectively. The degree of nutrient circulation is system specific, as it depends upon such factors as wind strength and duration, as well as lake or pool depth and productivity.

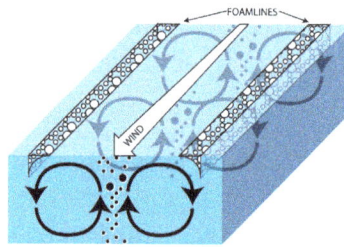

Fig. Illustration of Langmuir rotations; open circles=positively buoyant particles, closed circles=negatively buoyant particles

Chemistry

Oxygen is essential for organismal respiration. The amount of oxygen present in standing waters depends upon: 1) the area of transparent water exposed to the air, 2) the circulation of water within the system and 3) the amount of oxygen generated and used by organisms present. In shallow, plant-rich pools there may be great fluctuations of oxygen, with extremely high concentrations occurring during the day due to photosynthesis and very low values at night when respiration is the dominant process of primary producers. Thermal stratification in larger systems can also affect the amount of oxygen present in different zones. The epilimnion is oxygen rich because it circulates quickly, gaining oxygen via contact with the air. The hypolimnion, however, circulates very slowly and has no atmospheric contact. Additionally, fewer green plants exist in the hypolimnion, so there is less oxygen released from photosynthesis. In spring and fall when the epilimnion and hypolimnion mix, oxygen becomes more evenly distributed in the system. Low oxygen levels are characteristic of the profundal zone due to the accumulation of decaying vegetation and animal matter that "rains" down from the pelagic and benthic zones and the inability to support primary producers.

Phosphorus is important for all organisms because it is a component of DNA and RNA and is involved in cell metabolism as a component of ATP and ADP. Also, phosphorus is not found in large quantities in freshwater systems, limiting photosynthesis in primary producers, making it the main determinant of lentic system production. The phosphorus cycle is complex, but the model outlined below describes the basic pathways.

Phosphorus mainly enters a pond or lake through runoff from the watershed or by atmospheric deposition. Upon entering the system, a reactive form of phosphorus is usually taken up by algae and macrophytes, which release a non-reactive phosphorus compound as a byproduct of photosynthesis. This phosphorus can drift downwards and become part of the benthic or profundal sediment, or it can be remineralized to the reactive form by microbes in the water column. Similarly, non-reactive phosphorus in the sediment can be remineralized into the reactive form. Sediments are generally richer in phosphorus than lake water, however, indicating that this nutrient may have a long residency time there before it is remineralized and re-introduced to the system.

Lentic System Biota

Bacteria

Bacteria are present in all regions of lentic waters. Free-living forms are associated with decomposing organic material, biofilm on the surfaces of rocks and plants, suspended in the water column, and in the sediments of the benthic and profundal zones. Other forms are also associated with the guts of lentic animals as parasites or in commensal relationships. Bacteria play an important role in system metabolism through nutrient recycling.

Primary Producers

Nelumbo nucifera, an aquatic plant.

Algae, including both phytoplankton and periphyton are the principle photosynthe-sizers in ponds and lakes. Phytoplankton are found drifting in the water column of the pelagic zone. Many species have a higher density than water which should make them sink and end up in the benthos. To combat this, phytoplankton have developed den-sity changing mechanisms, by forming vacuoles and gas vesicles or by changing their shapes to induce drag, slowing their descent. A very sophisticated adaptation utilized by a small number of species is a tail-like flagellum that can adjust vertical position and allow movement in any direction. Phytoplankton can also maintain their presence in the water column by being circulated in Langmuir rotations. Periphytic algae, on the other hand, are attached to a substrate. In lakes and ponds, they can cover all benthic surfaces. Both types of plankton are important as food sources and as oxygen providers.

Aquatic plants live in both the benthic and pelagic zones and can be grouped according to their manner of growth: 1) emergent = rooted in the substrate but with leaves and flowers extending into the air, 2) floating-leaved = rooted in the substrate but with floating leaves, 3) submersed = growing beneath the surface and 4) free-floating macrophytes = not rooted in the substrate and floating on the surface. These various forms of macrophytes generally occur in different areas of the benthic zone, with emergent vegetation nearest the shoreline, then floating-leaved macrophytes, followed by submersed vegetation. Free-floating macrophytes can occur anywhere on the system's surface.

Aquatic plants are more buoyant than their terrestrial counterparts because freshwater has a higher density than air. This makes structural rigidity unimportant in lakes and ponds (except in the aerial stems and leaves). Thus, the leaves and stems of most aquatic plants use less energy to construct and maintain woody tissue, investing that energy into fast growth instead. In order to contend with stresses induced by wind and waves, plants must be both flexible and tough. Light, water depth and substrate types are the most important factors controlling the distribution of submerged aquatic plants. Macrophytes are sources of food, oxygen, and habitat structure in the benthic zone, but cannot penetrate the depths of the euphotic zone and hence are not found there.

Invertebrates

Water striders are predatory insects which rely on surface tension to walk on top of water. They live on the surface of ponds, marshes, and other quiet waters. They can move very quickly, up to 1.5 m/s.

Zooplankton are tiny animals suspended in the water column. Like phytoplankton, these species have developed mechanisms that keep them from sinking to deeper waters, including drag-inducing body forms and the active flicking of appendages such as antennae or spines. Remaining in the water column may have its advantages in terms of feeding, but this zone's lack of refugia leaves zooplankton vulnerable to predation. In response, some species, especially Daphnia sp., make daily vertical migrations in the water column by passively sinking to the darker lower depths during the day and actively moving towards the surface during the night. Also, because conditions in a lentic system can be quite variable across seasons, zooplankton have the ability to switch from laying regular eggs to resting eggs when there is a lack of food, temperatures fall below 2 °C, or if predator abundance is high. These resting eggs have a diapause, or dormancy period that should allow the zooplankton to encounter conditions that are more favorable to survival when they finally hatch. The invertebrates that inhabit the benthic zone are numerically dominated by small species and are species rich compared to the zooplankton of the open water. They include Crustaceans (e.g. crabs, crayfish, and shrimp), molluscs (e.g. clams and snails), and numerous types of insects. These organisms are mostly found in the areas of macrophyte growth, where the richest resources, highly oxygenated water, and warmest portion of the ecosystem are found. The structurally diverse macrophyte beds are important sites for the accumulation of organic matter, and provide an ideal area for colonization. The sediments and plants also offer a great deal of protection from predatory fishes.

Very few invertebrates are able to inhabit the cold, dark, and oxygen poor profundal zone. Those that can are often red in color due to the presence of large amounts of hemoglobin, which greatly increases the amount of oxygen carried to cells. Because the concentration of oxygen within this zone is low, most species construct tunnels or borrows in which they can hide and make the minimum movements necessary to circulate water through, drawing oxygen to them without expending much energy.

Fish and other Vertebrates

Fish have a range of physiological tolerances that are dependent upon which species they belong to. They have different lethal temperatures, dissolved oxygen requirements, and spawning needs that are based on their activity levels and behaviors. Because fish are highly mobile, they are able to deal with unsuitable abiotic factors in one zone by simply moving to another. A detrital feeder in the profundal zone, for example, that finds the oxygen concentration has dropped too low may feed closer to the benthic zone. A fish might also alter its residence during different parts of its life history: hatching in a sediment nest, then moving to the weedy benthic zone to develop in a protected environment with food resources, and finally into the pelagic zone as an adult.

Other vertebrate taxa inhabit lentic systems as well. These include amphibians (e.g. salamanders and frogs), reptiles (e.g. snakes, turtles, and alligators), and a large number of waterfowl species. Most of these vertebrates spend part of their time in terrestrial habitats and thus are not directly affected by abiotic factors in the lake or pond. Many fish species are important as consumers and as prey species to the larger vertebrates.

Trophic Relationships

Primary Producers

Lentic systems gain most of their energy from photosynthesis performed by aquatic plants and algae. This autochthonous process involves the combination of carbon dioxide, water, and solar energy to produce carbohydrates and dissolved oxygen. Within a lake or pond, the potential rate of photosynthesis generally decreases with depth due to light attenuation. Photosynthesis, however, is often low at the top few millimeters of the surface, likely due to inhibition by ultraviolet light. The exact depth and photosynthetic rate measurements of this curve are system specific and depend upon: 1) the total biomass of photosynthesizing cells, 2) the amount of light attenuating materials and 3) the abundance and frequency range of light absorbing pigments (i.e. chlorophylls) inside of photosynthesizing cells. The energy created by these primary producers is important for the community because it is transferred to higher trophic levels via consumption.

Bacteria

The vast majority of bacteria in lakes and ponds obtain their energy by decomposing vegetation and animal matter. In the pelagic zone, dead fish and the occasional allochthonous input of litterfall are examples of coarse particulate organic matter (CPOM>1 mm). Bacteria degrade these into fine particulate organic matter (FPOM<1 mm) and then further into usable nutrients. Small organisms such as plankton are also characterized as FPOM. Very low concentrations of nutrients are released during decomposition because the bacteria are utilizing them to build their own biomass. Bacteria, however, are consumed by protozoa, which are in turn consumed by zooplankton, and then further up the trophic levels. Nutrients, including those that contain carbon and phosphorus, are reintroduced into the water column at any number of points along this food chain via excretion or organism death, making them available again for bacteria. This regeneration cycle is known as the microbial loop and is a key component of lentic food webs.

The decomposition of organic materials can continue in the benthic and profundal zones if the matter falls through the water column before being completely digested by the pelagic bacteria. Bacteria are found in the greatest abundance here in sediments, where they are typically 2-1000 times more prevalent than in the water column.

Benthic Invertebrates

Benthic invertebrates, due to their high level of species richness, have many methods of prey capture. Filter feeders create currents via siphons or beating cilia, to pull water and its nutritional contents, towards themselves for straining. Grazers use scraping, rasping, and shredding adaptations to feed on periphytic algae and macrophytes. Members of the collector guild browse the sediments, picking out specific particles with rapto-rial appendages. Deposit feeding invertebrates indiscriminately consume sediment, digesting any organic material it contains. Finally, some invertebrates belong to the predator guild, capturing and consuming living animals. The profundal zone is home to a unique group of filter feeders that use small body movements to draw a current through burrows that they have created in the sediment. This mode of feeding requires the least amount of motion, allowing these species to conserve energy. A small number of invertebrate taxa are predators in the profundal zone. These species are likely from other regions and only come to these depths to feed. The vast majority of invertebrates in this zone are deposit feeders, getting their energy from the surrounding sediments.

Fish

Fish size, mobility, and sensory capabilities allow them to exploit a broad prey base, covering multiple zonation regions. Like invertebrates, fish feeding habits can be cat-egorized into guilds. In the pelagic zone, herbivores graze on periphyton and mac-rophytes or pick phytoplankton out of the water column. Carnivores include fishes that feed on zooplankton in the water column (zooplanktivores), insects at the water's surface, on benthic structures, or in the sediment (insectivores), and those that feed on other fish (piscivores). Fish that consume detritus and gain energy by processing its organic material are called detritivores. Omnivores ingest a wide variety of prey, encompassing floral, faunal, and detrital material. Finally, members of the parasitic guild acquire nutrition from a host species, usually another fish or large vertebrate. Fish taxa are flexible in their feeding roles, varying their diets with environmental conditions and prey availability. Many species also undergo a diet shift as they devel-op. Therefore, it is likely that any single fish occupies multiple feeding guilds within its lifetime.

Lentic Food Webs

The lentic biota are linked in complex web of trophic relationships. These organisms can be considered to loosely be associated with specific trophic groups (e.g. prima-ry producers, herbivores, primary carnivores, secondary carnivores, etc). Scientists have developed several theories in order to understand the mechanisms that control the abundance and diversity within these groups. Very generally, top-down processes dictate that the abundance of prey taxa is dependent upon the actions of consumers from higher trophic levels. Typically, these processes operate only between two trophic levels, with no effect on the others. In some cases, however, aquatic systems experi-ence a trophic cascade; for example, this might occur if primary producers experience less grazing by herbivores because these herbivores are suppressed by carnivores. Bot-tom-up processes are functioning when the abundance or diversity of members of high-er trophic levels is dependent upon the availability or quality of resources from lower levels. Finally, a combined regulating theory, bottom-up: top-down, combines the pre-dicted influences of consumers and resource availability. It predicts that trophic levels close to the lowest trophic levels will be most influenced by bottom-up forces, while top-down effects should be strongest at top levels.

Community Patterns and Diversity

Local Species Richness

The biodiversity of a lentic system increases with the surface area of the lake or pond. This is attributable to the higher likelihood of partly terrestrial species of finding a larger system. Also, because larger systems typically have larger populations, the chance of extinction is decreased. Additional factors, including temperature regime, pH, nutrient availability, habitat complexity, speciation rates, competition, and predation, have been linked to the number of species present within systems.

Succession Patterns in Plankton Communities – The PEG Model

Phytoplankton and zooplankton communities in lake systems undergo seasonal succession in relation to nutrient availability, predation, and competition. Sommer described these patterns as part of the Plankton Ecology Group (PEG) model, with 24 statements constructed from the analysis of numerous systems. The following includes a subset of these statements, as explained by Brönmark and Hansson illustrating succession through a single seasonal cycle:

Winter:

- Increased nutrient and light availability result in rapid phytoplankton growth towards the end of winter. The dominant species, such as diatoms, are small and have quick growth capabilities. These plankton are consumed by zooplankton, which become the dominant plankton taxa.

Spring:

- A clear water phase occurs, as phytoplankton populations become depleted due to increased predation by growing numbers of zooplankton.

Summer:

- Zooplankton abundance declines as a result of decreased phytoplankton prey and increased predation by juvenile fishes.

- With increased nutrient availability and decreased predation from zooplankton, a diverse phytoplankton community develops.

- As the summer continues, nutrients become depleted in a predictable order: phosphorus, silica, and then nitrogen. The abundance of various phytoplankton species varies in relation to their biological need for these nutrients.

- Small-sized zooplankton become the dominant type of zooplankton because they are less vulnerable to fish predation.

Fall:

- Predation by fishes is reduced due to lower temperatures and zooplankton of all sizes increase in number.

Winter:

- Cold temperatures and decreased light availability result in lower rates of primary production and decreased phytoplankton populations. 10. Reproduction in zooplankton decreases due to lower temperatures and less prey.

The PEG model presents an idealized version of this succession pattern, while natural systems are known for their variation.

Latitudinal Patterns

There is a well-documented global pattern that correlates decreasing plant and animal diversity with increasing latitude, that is to say, there are fewer species as one moves towards the poles. The cause of this pattern is one of the greatest puzzles for ecologists today. Theories for its explanation include energy availability, climatic variability, disturbance, competition, etc. Despite this global diversity gradient, this pattern can be weak for freshwater systems compared to global marine and terrestrial systems. This may be related to size, as Hillebrand and Azovsky found that smaller organisms (protozoa and plankton) did not follow the expected trend strongly, while larger species (vertebrates) did. They attributed this to better dispersal ability by smaller organisms, which may result in high distributions globally.

Natural Lake Lifecycles

Lake Creation

Lakes can be formed in a variety of ways. The oldest and largest systems are the result of tectonic activities. The rift lakes in Africa, for example are the result of seismic activity along the site of separation of two tectonic plates. Ice-formed lakes are created when glaciers recede, leaving behind abnormalities in the landscape shape that are then filled with water. Finally, oxbow lakes are fluvial in origin, resulting when a meandering river bend is pinched off from the main channel.

Natural Extinction

All lakes and ponds receive sediment inputs. Since these systems are not really expanding, it is logical to assume that they will become increasingly shallower in depth, eventually becoming wetlands or terrestrial vegetation. The length of this process should depend upon a combination of depth and sedimentation rate. Moss gives the example of Lake Tanganyika, which reaches a depth of 1500 m and has a sedimentation rate of 0.5 mm/yr. Assuming that sedimentation is not influenced by anthropogenic factors, this system should go extinct in approximately 3 million years. Shallow lentic systems might also fill in as swamps encroach inward from the edges. These processes operate on a much shorter timescale, taking hundreds to thousands of years to complete the extinction process.

Human Impacts

Acidification

Sulfur dioxide and nitrogen oxides are naturally released from volcanoes, organic compounds in the soil, wetlands, and marine systems, but the majority of these compounds come from the combustion of coal, oil, gasoline, and the smelting of ores containing sulfur. These substances dissolve in atmospheric moisture and enter lentic systems as acid rain. Lakes and ponds that contain bedrock that is rich in carbonates have a natural buffer, resulting in no alteration of pH. Systems without this bedrock, however, are very sensitive to acid inputs because they have a low neutralizing capacity, resulting in pH declines even with only small inputs of acid. At a pH of 5–6 algal species diversity and biomass decrease considerably, leading to an increase in water transparency – a characteristic feature of acidified lakes. As the pH continues lower, all fauna becomes less diverse. The most significant feature is the disruption of fish reproduction. Thus, the population is eventually composed of few, old individuals that eventually die and leave the systems without fishes. Acid rain has been especially harmful to lakes in Scandinavia, western Scotland, west Wales and the north eastern United States.

Eutrophication

Eutrophic systems contain a high concentration of phosphorus (~30 µg/L), nitrogen (~1500 µg/L), or both. Phosphorus enters lentic waters from sewage treatment effluents, discharge from raw sewage, or from runoff of farmland. Nitrogen mostly comes from agricultural fertilizers from runoff or leaching and subsequent groundwater flow. This increase in nutrients required for primary producers results in a massive increase of phytoplankton growth, termed a plankton bloom. This bloom decreases water transparency, leading to the loss of submerged plants. The resultant reduction in habitat structure has negative impacts on the species' that utilize it for spawning, maturation and general survival. Additionally, the large number of short-lived phytoplankton result in a massive amount of dead biomass settling into the sediment. Bacteria need large amounts of oxygen to decompose this material, reducing the oxygen concentration of the water. This is especially pronounced in stratified lakes when the thermocline prevents oxygen rich water from the surface to mix with lower levels. Low or anoxic conditions preclude the existence of many taxa that are not physiologically tolerant of these conditions.

Invasive Species

Invasive species have been introduced to lentic systems through both purposeful events (e.g. stocking game and food species) as well as unintentional events (e.g. in ballast water). These organisms can affect natives via competition for prey or habitat, predation, habitat alteration, hybridization, or the introduction of harmful diseases and parasites. With regard to native species, invaders may cause changes in size and age structure, distribution, density, population growth, and may even drive populations to extinction. Examples of prominent invaders of lentic systems include the zebra mussel and sea lamprey in the Great Lakes.

River Ecosystem

River ecosystems are flowing waters that drain the landscape, and include the biotic (living) interactions amongst plants, animals and micro-organisms, as well as abiotic (nonliving) physical and chemical interactions of its many parts. River ecosystems are part of larger watershed networks or catchments, where smaller headwater streams drain into mid-size streams, which progressively drain into larger river networks.

River ecosystems are prime examples of lotic ecosystems. *Lotic* refers to flowing water, from *lotus*, meaning washed. Lotic waters range from springs only a few centimeters wide to major rivers kilometers in width. Lotic ecosystems can be contrasted with lentic ecosystems, which involve relatively still terrestrial waters such as lakes, ponds, and wetlands. Together, these two ecosystems form the more general study area of freshwater or aquatic ecology.

The following unifying characteristics make the ecology of running waters unique among aquatic habitats.

- Flow is unidirectional.

- There is a state of continuous physical change.

- There is a high degree of spatial and temporal heterogeneity at all scales (microhabitats).

- Variability between lotic systems is quite high.

- The biota is specialized to live with flow conditions.

Abiotic Factors

The non living components of an ecosystem are called abiotic components. E.g stone, air, soil, etc.

Water Flow

A pensive Cooplacurripa River.

Rapids in Mount Robson Provincial Park.

Unidirectional water flow is the key factor in lotic systems influencing their ecology. Stream flow can be continuous or intermittent, though. Stream flow is the result of the summative inputs from groundwater, precipitation, and overland flow. Water flow can vary between systems, ranging from torrential rapids to slow backwaters that almost seem like lentic systems. The speed or velocity of the water flow of the water column can also vary within a system and is subject to chaotic turbulence, though water velocity tends to be highest in the middle part of the stream channel (known as the thalveg). This turbulence results in divergences of flow from the mean downslope flow vector as typified by eddy currents. The mean flow rate vector is based on variability of friction with the bottom or sides of the channel, sinuosity, obstructions, and the incline gradient. In addition, the amount of water input into the system from direct precipitation, snowmelt, and groundwater can affect flow rate. The amount of water in a stream is measured as discharge (volume per unit time). As water flows downstream, streams and rivers most often gain water volume, so at base flow (i.e., no storm input), smaller headwater streams have very low discharge, while larger rivers have much higher discharge. The "flow regime" of a river or stream includes the general patterns of discharge over annual or decadal time scales, and may capture seasonal changes in flow.

While water flow is strongly determined by slope, flowing waters can alter the general shape or direction of the stream bed, a characteristic also known as geomorphology. The profile of the river water column is made up of three primary actions: erosion, transport, and deposition. Rivers have been described as "the gutters down which run the ruins of continents". Rivers are continuously eroding, transporting, and depositing substrate, sediment, and organic material. The continuous movement of water and entrained material creates a variety of habitats, including riffles, glides, and pools.

Light

Light is important to lotic systems, because it provides the energy necessary to drive primary production via photosynthesis, and can also provide refuge for prey species in shadows it casts. The amount of light that a system receives can be related to a combination of internal and external stream variables. The area surrounding a small stream, for example, might be shaded by surrounding forests or by valley walls. Larger river systems tend to be wide so the influence of external variables is minimized, and the sun reaches the surface. These rivers also tend to be more turbulent, however, and particles in the water increasingly attenuate light as depth increases. Seasonal and diurnal factors might also play a role in light availability because the angle of incidence, the angle at which light strikes water can lead to light lost from reflection. Known as Beer's Law, the shallower the angle, the more light is reflected and the amount of solar radiation received declines logarithmically with depth. Additional influences on light availability include cloud cover, altitude, and geographic position.

Temperature

Castle Geyser, Yellowstone National Park.

A forest stream in the winter near Erzhausen.

Most lotic species are poikilotherms whose internal temperature varies with their environment, thus temperature is a key abiotic factor for them. Water can be heated or cooled through radiation at the surface and conduction to or from the air and surrounding substrate. Shallow streams are typically well mixed and maintain a relatively uniform temperature within an area. In deeper, slower moving water systems, however, a strong difference between the bottom and surface temperatures may develop. Spring fed systems have little variation as springs are typically from groundwater sources, which are often very close to ambient temperature. Many systems show strong diurnal fluctuations and seasonal variations are most extreme in arctic, desert and temperate systems. The amount of shading, climate and elevation can also influence the temperature of lotic systems.

Chemistry

Water chemistry in river ecosystems varies depending on which dissolved solutes and gases are present in the water column of the stream. Specifically river water can include, apart from the water itself,

- Dissolved inorganic matter and major ions (calcium, sodium, magnesium, potassium, bicarbonate, sulfide, chloride).

- Dissolved inorganic nutrients (nitrogen, phosphorus, silica).

- Suspended and dissolved organic matter.

- Gases (nitrogen, nitrous oxide, carbon dioxide, oxygen).

- Trace metals and pollutants.

Dissolved Ions and Nutrients

Dissolved stream solutes can be considered either *reactive* or *conservative*. Reactive solutes are readily biologically assimilated by the autotrophic and heterotrophic biota of the stream; examples can include inorganic nitrogen species such as nitrate or ammonium, some forms of phosphorus (e.g., soluble reactive phosphorus), and silica. Other solutes can be considered conservative, which indicates that the solute is not taken up and used biologically; chloride is often considered a conservative solute. Conservative solutes are often used as hydrologic tracers for water movement and transport. Both reactive and conservative stream water chemistry is foremost determined by inputs from the geology of its watershed, or catchment area. Stream water chemistry can also be influenced by precipitation, and the addition of pollutants from human sources. Large differences in chemistry do not usually exist within small lotic systems due to a high rate of mixing. In larger river systems, however, the concentrations of most nutrients, dissolved salts, and pH decrease as distance increases from the river's source.

Dissolved Gases

In terms of dissolved gases, oxygen is likely the most important chemical constituent of lotic systems, as all aerobic organisms require it for survival. It enters the water mostly via diffusion at the water-air interface. Oxygen's solubility in water decreases as water pH and temperature increases. Fast, turbulent streams expose more of the water's surface area to the air and tend to have low temperatures and thus more oxygen than slow, backwaters. Oxygen is a byproduct of photosynthesis, so systems with a high abundance of aquatic algae and plants may also have high concentrations of oxygen during the day. These levels can decrease significantly during the night when primary producers switch to respiration. Oxygen can be limiting if circulation between the surface and deeper layers is poor, if the activity of lotic animals is very high, or if there is a large amount of organic decay occurring.

Suspended Matter

Cascade in the Pyrénées.

Rivers can also transport suspended inorganic and organic matter. These materials can include sediment or terrestrially-derived organic matter that falls into the stream channel. Often, organic matter is processed within the stream via mechanical fragmentation, consumption and grazing by invertebrates, and microbial decomposition. Leaves and woody debris recognizable coarse particulate organic matter (CPOM) into particulate organic matter (POM), down to fine particulate organic matter. Woody and non-woody plants have different instream breakdown rates, with leafy plants or plant parts (e.g., flower petals) breaking down faster than woody logs or branches.

Substrate

The inorganic substrate of lotic systems is composed of the geologic material present in the catchment that is eroded, transported, sorted, and deposited by the current. Inorganic substrates are classified by size on the Wentworth scale, which ranges from boulders, to pebbles, to gravel, to sand, and to silt. Typically, substrate particle size decreases downstream with larger boulders and stones in more mountainous areas and sandy bottoms in lowland rivers. This is because the higher gradients of mountain streams facilitate a faster flow, moving smaller substrate materials further downstream for deposition. Substrate can also be organic and may include fine particles, autumn shed leaves, large woody debris such as submerged tree logs, moss, and semi-aquatic plants. Substrate deposition is not necessarily a permanent event, as it can be subject to large modifications during flooding events.

Biotic Factors

The living components of an ecosystem are called the biotic components. Streams have numerous types of biotic organisms that live in them, including bacteria, primary producers, insects and other invertebrates, as well as fish and other vertebrates.

Biofilm

Biofilm is the combination of algae, diatoms, fungi, bacteria, plankton, and other small microorganisms that exist in a film along the streambed or benthos. Biofilm assemblages themselves are complex, and add to the complexity of a streambed.

Bacteria

Bacteria are present in large numbers in lotic waters. Free-living forms are associated with decomposing organic material, biofilm on the surfaces of rocks and vegetation, in between particles that compose the substrate, and suspended in the water column. Other forms are also associated with the guts of lotic organisms as parasites or in commensal relationships. Bacteria play a large role in energy recycling.

Primary Producers

Periphyton.

Common water hyacinth in flower.

Algae, consisting of phytoplankton and periphyton, are the most significant sources of primary production in most streams and rivers. Phytoplankton float freely in the water column and thus are unable to maintain populations in fast flowing streams. They can, however, develop sizeable populations in slow moving rivers and backwaters. Periphyton are typically filamentous and tufted algae that can attach themselves to objects to avoid being washed away by fast currents. In places where flow rates are negligible or absent, periphyton may form a gelatinous, unanchored floating mat.

Plants exhibit limited adaptations to fast flow and are most successful in reduced currents. More primitive plants, such as mosses and liverworts attach themselves to solid objects. This typically occurs in colder headwaters where the mostly rocky substrate offers attachment sites. Some plants are free floating at the water's surface in dense mats like duckweed or water hyacinth. Others are rooted and may be classified as submerged or emergent. Rooted plants usually occur in areas of slackened current where fine-grained soils are found. These rooted plants are flexible, with elongated leaves that offer minimal resistance to current.

Living in flowing water can be beneficial to plants and algae because the current is usually well aerated and it provides a continuous supply of nutrients. These organisms are limited by flow, light, water chemistry, substrate, and grazing pressure. Algae and plants are important to lotic systems as sources of energy, for forming microhabitats that shelter other fauna from predators and the current, and as a food resource.

Insects and other Invertebrates

Up to 90% of invertebrates in some lotic systems are insects. These species exhibit tremendous diversity and can be found occupying almost every available habitat, including the surfaces of stones, deep below the substratum in the hyporheic zone, adrift in the current, and in the surface film.

Insects have developed several strategies for living in the diverse flows of lotic systems. Some avoid high current areas, inhabiting the substratum or the sheltered side of rocks. Others have flat bodies to reduce the drag forces they experience from living in running water. Some insects, like the giant water bug (Belostomatidae), avoid flood events by leaving the stream when they sense rainfall. In addition to these behaviors and body shapes, insects have different life history adaptations to cope with the naturally-occurring physical harshness of stream environments. Some insects time their life events based on when floods and droughts occur. For example, some mayflies synchronize when they emerge as flying adults with when snowmelt flooding usually occurs in Colorado streams. Other insects do not have a flying stage and spend their entire life cycle in the river.

Like most of the primary consumers, lotic invertebrates often rely heavily on the current to bring them food and oxygen. Invertebrates are important as both consumers and prey items in lotic systems.

The common orders of insects that are found in river ecosystems include Ephemeroptera (also known as a mayfly), Trichoptera (also known as a caddisfly), Plecoptera (also known as a stonefly, Diptera (also known as a true fly), some types of Coleoptera (also known as a beetle), Odonata (the group that includes the dragonfly and the damselfly), and some types of Hemiptera (also known as true bugs).

Additional invertebrate taxa common to flowing waters include mollusks such as snails, limpets, clams, mussels, as well as crustaceans like crayfish, amphipoda and crabs.

Fish and other Vertebrates

The brook trout is native to small streams, creeks, lakes, and spring ponds.

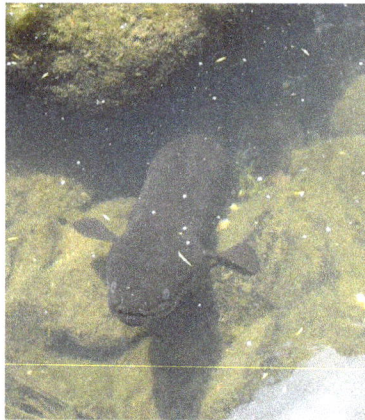

New Zealand longfin eels can weigh over 50 kilograms.

Fish are probably the best-known inhabitants of lotic systems. The ability of a fish species to live in flowing waters depends upon the speed at which it can swim and the duration that its speed can be maintained. This ability can vary greatly between species and is tied to the habitat in which it can survive. Continuous swimming expends a tremendous amount of energy and, therefore, fishes spend only short periods in full current. Instead, individuals remain close to the bottom or the banks, behind obstacles, and sheltered from the current, swimming in the current only to feed or change locations. Some species have adapted to living only on the system bottom, never venturing into the open water flow. These fishes are dorso-ventrally flattened to reduce flow resistance and often have eyes on top of their heads to observe what is happening above them. Some also have sensory barrels positioned under the head to assist in the testing of substratum.

Lotic systems typically connect to each other, forming a path to the ocean (spring → stream → river → ocean), and many fishes have life cycles that require stages in both fresh and salt water. Salmon, for example, are anadromous species that are born in freshwater but spend most of their adult life in the ocean, returning to fresh water only to spawn. Eels are catadromous species that do the opposite, living in freshwater as adults but migrating to the ocean to spawn.

Other vertebrate taxa that inhabit lotic systems include amphibians, such as salamanders, reptiles (e.g. snakes, turtles, crocodiles and alligators) various bird species, and mammals (e.g., otters, beavers, hippos, and river dolphins). With the exception of a few species, these vertebrates are not tied to water as fishes are, and spend part of their time in terrestrial habitats. Many fish species are important as consumers and as prey species to the larger vertebrates.

Trophic Relationships

Energy Inputs

1 cm

Pondweed is an autochthonous energy source.

Leaf litter is an allochthonous energy source.

Energy Sources can be Autochthonous or Allochthonous

- Autochthonous energy sources are those derived from within the lotic system. During photosynthesis, for example, primary producers form organic carbon compounds out of carbon dioxide and inorganic matter. The energy they produce is important for the community because it may be transferred to higher trophic levels via consumption. Additionally, high rates of primary production can introduce dissolved organic matter (DOM) to the waters. Another form of autochthonous energy comes from the decomposition of dead organisms and feces that originate within the lotic system. In this case, bacteria decompose the detritus or coarse particulate organic material (CPOM; >1 mm pieces) into fine particulate organic matter (FPOM; <1 mm pieces) and then further into inorganic compounds that are required for photosynthesis.

- Allochthonous energy sources are those derived from outside the lotic system, that is, from the terrestrial environment. Leaves, twigs, fruits, etc. are typical forms of terrestrial CPOM that have entered the water by direct litter fall or lateral leaf blow. In addition, terrestrial animal-derived materials, such as feces or carcasses that have been added to the system are examples of allochthonous CPOM. The CPOM undergoes a specific process of degradation. Allan gives the example of a leaf fallen into a stream. First, the soluble chemicals are dissolved and leached from the leaf upon its saturation with water. This adds to the DOM load in the system. Next microbes such as bacteria and fungi colonize the leaf, softening it as the mycelium of the fungus grows into it. The composition of the microbial community is influenced by the species of tree from which the leaves are shed (Rubbo and Kiesecker 2004). This combination of bacteria, fungi, and leaf are a food source for shredding invertebrates, which leave only FPOM after consumption. These fine particles may be colonized by microbes again or serve as a food source for animals that consume FPOM. Organic matter can also enter the lotic system already in the FPOM stage by wind, surface runoff, bank erosion, or groundwater. Similarly, DOM can be introduced through canopy drip from rain or from surface flows.

Invertebrates

Invertebrates can be organized into many feeding guilds in lotic systems. Some species are shredders, which use large and powerful mouth parts to feed on non-woody CPOM and their associated microorganisms. Others are suspension feeders, which use their setae, filtering aparati, nets, or even secretions to collect FPOM and microbes from the water. These species may be passive collectors, utilizing the natural flow of the system, or they may generate their own current to draw water, and also, FPOM in Allan. Members of the gatherer-collector guild actively search for FPOM under rocks and in other places where the stream flow has slackened enough to allow deposition. Grazing invertebrates utilize scraping, rasping, and browsing adaptations to feed on periphyton and detritus. Finally, several families are predatory, capturing and consuming animal prey. Both the number of species and the abundance of individuals within each guild is largely dependent upon food availability. Thus, these values may vary across both seasons and systems.

Fish

Fish can also be placed into feeding guilds. Planktivores pick plankton out of the water column. Herbivore-detritivores are bottom-feeding species that ingest both periphyton and detritus indiscriminately. Surface and water column feeders capture surface prey (mainly terrestrial and emerging insects) and drift (benthic invertebrates floating downstream). Benthic invertebrate feeders prey primarily on immature insects, but will also consume other benthic invertebrates. Top predators consume fishes and large invertebrates. Omnivores ingest a wide range of prey. These can be floral, faunal, and detrital in nature. Finally, parasites live off of host species, typically other fishes. Fish are flexible in their feeding roles, capturing different prey with regard to seasonal availability and their own developmental stage. Thus, they may occupy multiple feeding guilds in their lifetime. The number of species in each guild can vary greatly between systems, with temperate warm water streams having the most benthic invertebrate feeders, and tropical systems having large numbers of detritus feeders due to high rates of allochthonous input.

Community Patterns and Diversity

Iguazu Falls - An extreme lotic environment.

Beaver Run - A placid lotic environment.

Local Species Richness

Large rivers have comparatively more species than small streams. Many relate this pattern to the greater area and volume of larger systems, as well as an increase in habitat diversity. Some systems, however, show a poor fit between system size and species richness. In these cases, a combination of factors such as historical rates of speciation and extinction, type of substrate, microhabitat availability, water chemistry, temperature, and disturbance such as flooding seem to be important.

Resource Partitioning

Although many alternate theories have been postulated for the ability of guild-mates to coexist, resource partitioning has been well documented in lotic systems as a means of reducing competition. The three main types of resource partitioning include habitat, dietary, and temporal segregation.

Habitat segregation was found to be the most common type of resource partitioning in natural systems. In lotic systems, microhabitats provide a level of physical complexity that can support a diverse array of organisms. The separation of species by substrate preferences has been well documented for invertebrates. Ward was able to divide substrate dwellers into six broad assemblages, including those that live in: coarse substrate, gravel, sand, mud, woody debris, and those associated with plants, showing one layer of segregation. On a smaller scale, further habitat partitioning can occur on or around a single substrate, such as a piece of gravel. Some invertebrates prefer the high flow areas on the exposed top of the gravel, while others reside in the crevices between one piece of gravel and the next, while still others live on the bottom of this gravel piece.

Dietary segregation is the second-most common type of resource partitioning. High degrees of morphological specializations or behavioral differences allow organisms to use specific resources. The size of nets built by some species of invertebrate suspension feeders, for example, can filter varying particle size of FPOM from the water. Similarly, members in the grazing guild can specialize in the harvesting of algae or detritus depending upon the morphology of their scraping apparatus. In addition, certain species seem to show a preference for specific algal species.

Temporal segregation is a less common form of resource partitioning, but it is nonetheless an observed phenomenon. Typically, it accounts for coexistence by relating it to differences in life history patterns and the timing of maximum growth among guild mates. Tropical fishes in Borneo, for example, have shifted to shorter life spans in response to the ecological niche reduction felt with increasing levels of species richness in their ecosystem.

Persistence and Succession

Over long time scales, there is a tendency for species composition in pristine systems to remain in a stable state. This has been found for both invertebrate and fish species. On shorter time scales, however, flow variability and unusual precipitation patterns decrease habitat stability and can all lead to declines in persistence levels. The ability to maintain this persistence over long time scales is related to the ability of lotic systems to return to the original community configuration relatively quickly after a disturbance. This is one example of temporal succession, a site-specific change in a community involving changes in species composition over time. Another form of temporal succession might occur when a new habitat is opened up for colonization. In these cases, an entirely new community that is well adapted to the conditions found in this new area can establish itself.

River Continuum Concept

↑ Meandering stream in Waitomo.

River Gryffe.

Rocky stream.

The River continuum concept (RCC) was an attempt to construct a single framework to describe the function of temperate lotic ecosystems from the headwaters to larger rivers and relate key characteristics to changes in the biotic community. The physical basis for RCC is size and location along the gradient from a small stream eventually linked to a large river. Stream order is used as the physical measure of the position along the RCC.

According to the RCC, low ordered sites are small shaded streams where allochthonous inputs of CPOM are a necessary resource for consumers. As the river widens at mid-ordered sites, energy inputs should change. Ample sunlight should reach the bottom in these systems to support significant periphyton production. Additionally, the biological processing of CPOM (Coarse Particulate Organic Matter - larger than 1 mm) inputs at upstream sites is expected to result in the transport of large amounts of FPOM (Fine Particulate Organic Matter - smaller than 1 mm) to these downstream ecosystems. Plants should become more abundant at edges of the river with increasing river size, especially in lowland rivers where finer sediments have been deposited and facilitate rooting. The main channels likely have too much current and turbidity and a lack of substrate to support plants or periphyton. Phytoplankton should produce the only autochthonous inputs here, but photosynthetic rates will be limited due to turbidity and mixing. Thus, allochthonous inputs are expected to be the primary energy source for large rivers. This FPOM will come from both upstream sites via the decomposition process and through lateral inputs from floodplains.

Biota should change with this change in energy from the headwaters to the mouth of these systems. Namely, shredders should prosper in low-ordered systems and grazers in mid-ordered sites. Microbial decomposition should play the largest role in energy production for low-ordered sites and large rivers, while photosynthesis, in addition to degraded allochthonous inputs from upstream will be essential in mid-ordered systems. As mid-ordered sites will theoretically receive the largest variety of energy inputs, they might be expected to host the most biological diversity.

Just how well the RCC actually reflects patterns in natural systems is uncertain and its generality can be a handicap when applied to diverse and specific situations. The most noted criticisms of the RCC are: 1. It focuses mostly on macroinvertebrates, disregarding that plankton and fish diversity is highest in high orders; 2. It relies heavily on the fact that low ordered sites have high CPOM inputs, even though many streams lack riparian habitats; 3. It is based on pristine systems, which rarely exist today; and 4. It is centered around the functioning of temperate streams. Despite its shortcomings, the RCC remains a useful idea for describing how the patterns of ecological functions in a lotic system can vary from the source to the mouth.

Disturbances such as congestion by dams or natural events such as shore flooding are not included in the RCC model. Various researchers have since expanded the model to account for such irregularities. For example, J.V. Ward and J.A. Stanford came up with the Serial Discontinuity Concept in 1983, which addresses the impact of geomorphologic disorders such as congestion and integrated inflows. The same authors presented the Hyporheic Corridor concept in 1993, in which the vertical (in depth) and lateral (from shore to shore) structural complexity of the river were connected. The flood pulse concept, developed by W.J. Junk in 1989, further modified by P.B. Bayley in 1990 and K. Tockner in 2000, takes into account the large amount of nutrients and organic material that makes its way into a river from the sediment of surrounding flooded land.

Human Impacts

Pollution

River pollution can include but is not limited to: increasing sediment export, excess nutrients from fertilizer or urban runoff, sewage and septic inputs, plastic pollution, nano-particles, pharmaceuticals and personal care products, synthetic chemicals, road salt, inorganic contaminants (e.g., heavy metals), and even heat via thermal pollutions. The effects of pollution often depend on the context and material, but can reduce ecosystem functioning, limit ecosystem services, reduce stream biodiversity, and impact human health.

Pollutant sources of lotic systems are hard to control because they can derive, often in small amounts, over a very wide area and enter the system at many locations along its length. While direct pollution of lotic systems has been greatly reduced in the United States under the government's Clean Water Act, contaminants from diffuse non-point sources remain a large problem. Agricultural fields often deliver large quantities of sediments, nutrients, and chemicals to nearby streams and rivers. Urban and residential areas can also add to this pollution when contaminants are accumulated on impervious surfaces such as roads and parking lots that then drain into the system. Elevated nutrient concentrations, especially nitrogen and phosphorus which are key components of fertilizers, can increase periphyton growth, which can be particularly dangerous in slow-moving streams. Another pollutant, acid rain, forms from sulfur dioxide and nitrous oxide emitted from factories and power stations. These substances readily dissolve in atmospheric moisture and enter lotic systems through precipitation. This can lower the pH of these sites, affecting all trophic levels from algae to vertebrates. Mean species richness and total species numbers within a system decrease with decreasing pH.

Flow Modification

A weir on the River Calder.

Flow modification can occur as a result of dams, water regulation and extraction, channel modification, and the destruction of the river floodplain and adjacent riparian zones.

Dams alter the flow, temperature, and sediment regime of lotic systems. Additionally, many rivers are dammed at multiple locations, amplifying the impact. Dams can cause enhanced clarity and reduced variability in stream flow, which in turn cause an increase in periphyton abundance. Invertebrates immediately below a dam can show reductions in species richness due to an overall reduction in habitat heterogeneity. Also, thermal changes can affect insect development, with abnormally warm winter temperatures obscuring cues to break egg diapause and overly cool summer temperatures leaving too few acceptable days to complete growth. Finally, dams fragment river systems, isolating previously continuous populations, and preventing the migrations of anadromous and catadromous species.

Invasive Species

Invasive species have been introduced to lotic systems through both purposeful events (e.g. stocking game and food species) as well as unintentional events (e.g. hitchhikers on boats or fishing waders). These organisms can affect natives via competition for prey or habitat, predation, habitat alteration, hybridization, or the introduction of harmful diseases and parasites. Once established, these species can be difficult to control or eradicate, particularly because of the connectivity of lotic systems. Invasive species can be especially harmful in areas that have endangered biota, such as mussels in the Southeast United States, or those that have localized endemic species, like lotic systems west of the Rocky Mountains, where many species evolved in isolation.

Marine Ecosystem

Marine ecosystems can be defined as the interaction of plants, animals, and the marine environment. The seas and oceans are rich in animal life, from whales and fish to starfish and sponges. The nature of marine communities depends upon depth, with the surface and the ocean floor being particularly rich in biodiversity. Similar communities are often found at similar depth, even though they may be widely separated geographically. There are also complex marine habitats, such as mangrove swamps, estuaries, and coral reefs, occurring near the shore.

Marine ecosystems depend largely upon phytoplankton, which are photosynthetic algae living near the surface of the water where the sun penetrates. Tiny herbivores feed on the phytoplankton and these, in turn, are eaten by increasingly larger animals, ending with larger fish and sharks at the top of the marine food web. The seas and oceans are important for humans as a resource for fish and other marine products, but its ecosystems are threatened by exploitation, such as overfishing and pollution.

Until the middle of the nineteenth century, it was assumed that few, if any, animals and plants lived in the seas and oceans because the waters were dark and cold. The first clues to the existence of rich and complex marine ecosystems came from broken underwater telegraph wires. When these were retrieved, various unusual and previously unknown creatures were found clinging onto them. The *HMS Challenger*, a British naval vessel, carried out the first-ever oceanographic survey between 1872 and 1876, exploring as deep as 18,700 ft (5,700 m) in the Pacific Ocean. The expedition returned with thousands of specimens, many of which were previously unknown to science.

The marine environment is where life evolved in the first place. Seawater contains sodium chloride (NaCl) and other mineral salts, and has remained at roughly the same salt concentration for millions of years. Salinity, as the salt concentration is called, is around one ounce per liter, of which 90% is composed of sodium chloride. This happens to be the same sodium chloride concentration as living cells, making it a natural environment for organisms to deal with.

Water covers around 70% of Earth's surface. Its average depth is just over 2 mi (3.2 km)—ranging from a few inches close to the shore to around 7 mi (11.2 km) at its greatest depth. The oceans alone provide more than 170 times more living space than land, air, and freshwater put together. The weight of water does exert pressure upon those organisms living there. This pressure will increase by one atmosphere for every 33 ft (10 m) of depth. But, like organisms living with atmospheric pressure on land, deep-sea animals such as fish and snails have the same pressure inside and outside their bodies. Most sea creatures are composed mainly of water and, since liquids are incompressible, they do not experience adverse effects on moving from one depth to another.

The seas are constantly in motion due to surface currents and deeper circulation currents, which means that cold salty water, which is heavier, sinks and is replaced by warmer, less salty water. This mixes the water, making its chemistry uniform as well as carrying oxygen (O) from the surface to deeper layers, making life down there possible. The temperature of surface water varies from 104 °F (40 °C) in tropical waters to about 35 °F (1.9 °C) for seawater in the Arctic and Antarctic. The depths of the oceans are always very cold, even in tropical regions, at 32 to 37 °F (0 to 3 °C).

Compared to the land, the marine environment is not so rich in different species. Only one tenth of the nearly 2 million animal species known are found in the sea, and only around 4,000 plants, compared to a quarter of a million on land. But marine ecosystems tend to be more diverse, with 28 different phyla existing in the oceans, compared to only 11 on land.

Temperature, salinity, and the availability of oxygen, light, and nutrients shape the marine ecosystems. Thousands of different invertebrates make their home in the ocean, often providing food for larger species. For instance, there are about 5,000 species of sponges, of a wide variety of colors and shapes. These animals have the least complex body structure of all multi-celled creatures, consisting of an outer layer of tissue and an innerlayer of either silica (SiO_2) or calcium carbonate ($CaCO_3$). The echinoderms, which include starfish, sea urchins, and sea cucumbers, are a group of exclusively marine bottom-dwelling invertebrates, characterized by hard, spiny skin. Around 6,000 species dwell in the world's salt waters, often in rocky pools around beaches.

The cnidarians are another important marine invertebrate group and include the jellyfish, corals, and sea anemones. These organisms are characterized by their soft, watery bodies. The molluscs, which are found on land and sea, include the gastropods (limpets, slugs, and snails), cephalopods (octopus, squid, and cuttlefish), and bivalves (clams, mussels, cockles, oysters, and scallops). Most crustaceans, of which there are about 39,000 species, are marine and include crabs, lobsters, shrimps, barnacles, and woodlice. The copepods, which are tiny crustaceans ranging from 0.02 to 0.7 in (0.05 to 1.8 cm) in length are particularly abundant. There could be hundreds of thousands of copepods in a cubic meter of surface water and they are an important source of food for predators like marine worms and the smaller jellyfish.

Marine worms are another diverse group, many of which are completely different from the well-known earthworm. For instance, the arrow worm is a major predator of the copepods. Only 1-in (2.5-cm) long, it is also known as the chaetognath worm, which means "bristle jaw". The worm darts and grabs its prey in its jaws. The nematodes, or thread worms, are among the tiniest of creatures in the marine environment, being just a fraction of an inch long. They live in sediments and feed off bacteria. As a group, the nematode worms have not been studied much, and researchers believe there may be thousands of species remaining to be discovered. At the other extreme are the tube worms found on the ocean floor, which may be up to 6 ft (1.8 m) in length. These creatures are so unusual that they have been placed in a phylum of their own. They have no mouths, but their bodies are filled with chemosynthetic bacteria, which extract energy from minerals rather than sunlight, and these provide much of their food supply.

Birds, mammals, and fish all live in or around the marine environment, but fish are the only vertebrates that are purely aquatic and are found in both freshwater and saltwater around the world. Around half of the 25,000 known species of fish live in the marine environment, mainly in shallower, warmer waters. Around 1,000 fish species occupy the open ocean. Fish that live in the deep ocean are generally black, brown, or gray, without the silvery camouflage that characterizes those living nearer the surface. Some are buoyant and swim up and down the depths of the water, searching for food, while others, including sharks and rays, are heavier and sink if they cease swimming. They tend to stay in place, catching food as it passes, or making just short hunting excursions.

Whales are marine mammals, breathing with lungs, and the largest animals in the oceans. There are two suborders: the baleen whales, which do not have teeth and filter feed on massive amounts of plankton; and the toothed whales, which feed on fish and squid. Toothed whales have the remarkable ability to navigate through the ocean by echolocation. The whales, sharks, and giant squid represent the top end of the marine ecosystem. Their sheer size has made them famous in marine culture and folklore.

In the open ocean, the ecosystems are vertically stratified. That is, different plants and animals are found at different depths. Sunlight penetrates to a depth of only 3 ft (0.9 m) or so in the cloudy waters of an estuary, compared to up to 300 ft (90 m) in the clearer waters of the open ocean. In this relatively light region, known as the euphotic zone, there is a net primary production of food by photosynthesis carried out by 4,000 or so species of phytoplankton.

The column of water extending down from the surface to a depth of about 2.5 mi (4 km) is called the pelagic zone and is composed of the epipelagic (top), mesopelagic (middle or twilight), and bathypelagic (bottom) zones. Below this, extending to about 4 mi (6.4 km), are the abyssal and hadal zones. There is no light in these zones close to the ocean floor other than what is emitted from the organisms themselves. Thousands of species, including bacteria, squid, and fish, emit flashes of light by a process known as bioluminesence to distract predators and locate prey.

The oceans and seas contain some specific ecological niches. For instance parts of the ocean floor, or benthos, contain hydrothermal vents, where jets of water containing sulfur (S) compounds gush up, heated by the magma beneath the ocean floor. The ecological communities of microbes, worms, and mussels living around hydrothermal vents were known only since 1977. They can withstand temperatures of up to 660 °F (350 °C), and the microbes are capable of extracting biochemical energy from the sulfur compounds.

Shallow areas of the seas and oceans have rich ecosystems too. The sea floor slopes gradually from the shore out to the deeper ocean. The area near the shore is called the littoral zone, and many fish and shellfish are found here. Filter feeders such as mussels and barnacles live in rocky-bottomed shores, while soft-bottomed beaches are home to scavengers like shrimps and polychaete worms. The number of species increases nearer to water. On sand that is never completely submerged, it is necessary to dig in to find worms and sand crabs, while sea anemones and sea cucumbers are found near rocky shores. Estuaries, where river meets the sea, mixes freshwater and mud sediments with seawater and are often rich in fish because the nutrient levels in the water are high.

Mangrove swamps or forests are another important area near the tropical shores. Mangroves are salt-resistant trees that support fish or shrimp. Coral reefs such as the Great Barrier Reef, which stretches for around 1,200 mi (1,930 km) along the eastern coast of Australia, are probably one of the best-known marine ecosystems. They consist of animals called stony corals, each of which is a polyp with tentacles that can trap organisms. The polyps live in symbiosis with photosynthetic algae. The algae provide them with food, and the polyps provide the algae with protection. Each polyp lies in a shell of calcium carbonate and they are joined together by a sheet of tissue to form a giant organism and ecosystem in its own right. Coral reefs tend to occur in warm, shallow, clear waters where they provide a home for a diverse community of fish, worms, and crustaceans, protecting the small fish from larger predatory fish.

The marine food web is based upon the photosynthetic activity of phytoplankton, which is often greatest near the shore, where the water tends to be richer in nutrients like nitrogen (N) and phosphorus (P) because of runoff from land. Many tiny herbivores feed directly on phytoplankton, as well as the juvenile stages of certain squid and fish. These miniature predators will, in turn, be meals for slightly bigger creatures. The pattern goes on, up to the larger fish and sharks that are the equivalent of wolves and lions on the land. Generally, animals do not feed on those that are more than a tenth their size. However, there are exceptions to this. Some sharks will feed on animals as big or bigger than themselves, taking bite-sized lumps out of them. Fish with wide mouths and big teeth in the deeper layers can swallow bigger animals. The blue whale, the largest animal in the ocean, dines solely on tiny shrimp called krill. Meanwhile, dead phytoplankton and other organisms sink to the bottom of the ocean, forming a "marine snow" that benthic organisms, such as crabs and fish, depend upon as a food source.

Impacts and Issues

Human activity impacts upon marine ecosystems just as it does on land, with the areas nearest the coast being most affected. For instance, the coral reefs are among the world's most threatened biomes. They are affected by a range of factors, including destructive fishing practices, pollution, and sewage. Meanwhile, global warming is already causing bleaching of coral reefs. Increased temperatures destroy the symbiotic relationship between the coral and the algae that live there. According to research from the United Nations, one third of coral reefs around the world are destroyed already, 60% are damaged and will likely be dead by 2030. Mangrove swamps are similarly at risk.

Estuaries and shore areas are also at risk from pollution, which can cause eutrophication by raising nutrient levels in the water. Eutrophication encourages the growth of decomposers that consume available oxygen in the water. As oxygen levels fall, fish and other organisms begin to die off. Eutrophication also causes an overgrowth of algae in the water, often visible as a red, yellow, or green scum on the surface and a visible sign of a threatened ecosystem. These dead zones are found in many areas around the world, such as the Mediterranean and the East Coast of the United States.

The deep ocean has also been used as a dump for low-level radioactive waste, although this was banned in 1993. There have also been discussions on burying medium and high-level radioactive waste from nuclear power stations. If land alternatives prove too risky, these options may be put into practice, but no one knows what the long-term impact might be upon marine ecosystems, however secure the waste was made.

Humans do not just put things into the seas and oceans. They take things out as well. Fishing is a traditional activity, with fish being an important protein source in the human diet. However, global fish harvesting has increased 5 times during the last 50 years or so, partly because fishing technology has become more efficient and partly because of the increase in human population, which has increased demand for food. The oceans can probably support a fish harvest of about 100 million tons (90 million metric tons) of fish caught per year. As the fishing industry has expanded, these limits are being reached. Fishing vessels now have to travel farther and farther to get catches. Not only does this hurt the economy, there could also be as-yet-unknown disturbances to the marine ecosystem by driving fish stocks down in this way.

People are also seeking to exploit the ocean for oil and gas. Drilling offshore began in 1947 in the Gulf of Mexico, and now there are thousands of such developments. It may be that efforts to extract oil and gas will go deeper still, despite the difficulties of the technology, with unknown effects on the marine environment. There has also been discussion about whether it might be possible to exploit the sea bed as a source of minerals. As the world's population grows and industrial development spreads, the pressure to use the ocean as a resource can only increase, with unknown impacts on marine ecosystems.

References

- Aquatic-ecosystems, energy-government-and-defense-magazines, environment: encyclopedia. com, Retrieved 30 April, 2019

- G., wetzel, robert (2001). Limnology : lake and river ecosystems (3rd ed.). San diego: academic press. Isbn 978-0127447605. Oclc 46393244

- Xu, f (september 2001). "lake ecosystem health assessment: indicators and methods". Water research. 35 (13): 3157–3167. Doi:10.1016/s0043-1354(01)00040-9. Issn 0043-1354

- Keddy, p.a. (2010). Wetland ecology: principles and conservation (2nd edition). Cambridge university press, cambridge, uk. Isbn 0521739675

- Hillebrand, h.; a. I. Azovsky (2001). "body size determines the strength of the latitudinal diversity gradient". Ecography. 24 (3): 251–256. Doi:10.1034/j.1600-0587.2001.240302.x

- Marine-ecosystems, energy-government-and-defense-magazines, environment: encyclopedia. com, Retrieved 29 June, 2019

Systems Ecology 7

- **Ecological Systems Theory**
- **The Ecological Laws of Thermodynamics**
- **Energy Flow in Ecosystem**
- **Nutrient Cycle**
- **Food Chain**
- **Food Web**
- **Ecological Pyramid**
- **Ecological Buffers**

Systems ecology is referred to the field of ecology that deals with the study of ecological systems. It focuses on the study of the ecological pyramid, food web, food chain, nutrient cycle, ecological buffers, energy flow, ecological laws, etc. All these diverse aspects related to systems ecology have been carefully analyzed in this chapter.

Systems ecology is a branch of ecosystem ecology (the study of energy budgets, biogeochemical cycles, and feeding and behavioral aspects of ecological communities) that attempts to clarify the structure and function of ecosystems by means of applied mathematics, mathematical models, and computer programs.

Ecological Systems Theory

Ecological systems theory (also called development in context or human ecology theory) offers a framework through which community psychologists examine individuals' relationships within communities and the wider society. The theory is also commonly referred to as the ecological/systems framework. It identifies five environmental systems with which an individual interacts. The Ecological systems theory was developed by Urie Bronfenbrenner.

The Five Systems

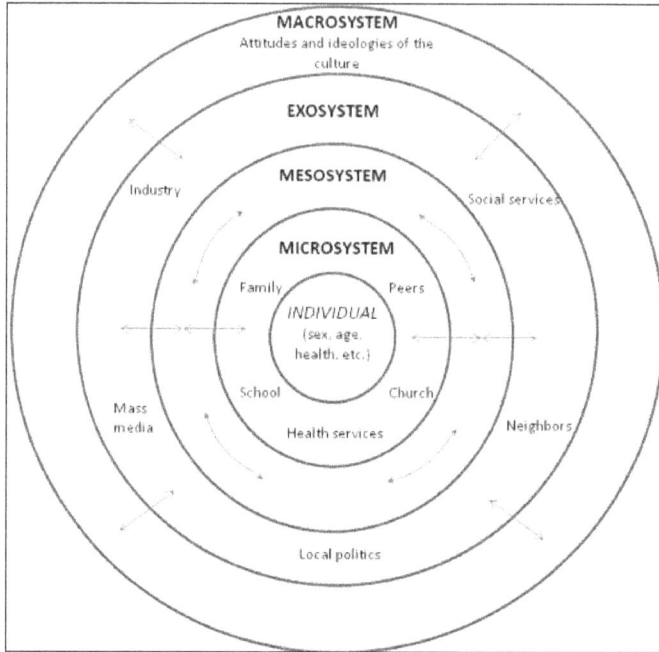

MACROSYSTEM
Attitudes and ideologies of the culture

EXOSYSTEM

MESOSYSTEM

MICROSYSTEM

Family   Peers

INDIVIDUAL
(sex, age, health, etc.)

Industry   Social services

School   Church

Mass media   Health services   Neighbors

Local politics

Bronfenbrenner's ecological systems theory.

Microsystem: Refers to the institutions and groups that most immediately and directly impact the child's development including: family, school, religious institutions, neighborhood, and peers.

Mesosystem: Interconnections between the microsystems, Interactions between the family and teachers, Relationship between the child's peers and the family

Exosystem: Involves links between a social setting in which the individual does not have an active role and the individual's immediate context. For example, a parent's or child's experience at home may be influenced by the other parent's experiences at work. The parent might receive a promotion that requires more travel, which might increase conflict with the other parent and change patterns of interaction with the child.

Macrosystem: Describes the culture in which individuals live. Cultural contexts include developing and industrialized countries, socioeconomic status, poverty, and ethnicity. A child, his or her parent, his or her school, and his or her parent's workplace are all part of a large cultural context. Members of a cultural group share a common identity, heritage, and values. The macrosystem evolves over time, because each successive generation may change the macrosystem, leading to their development in a unique macrosystem.

Chronosystem: The patterning of environmental events and transitions over the life course, as well as sociohistorical circumstances. For example, divorces are one transition. Researchers have found that the negative effects of divorce on children often peak in the first year after the divorce. By two years after the divorce, family interaction is less chaotic and more stable. An example of sociohistorical circumstances is the increase in opportunities for women to pursue a career during the last thirty years.

The person's own biology may be considered part of the microsystem; thus the theory has recently sometimes been called Bioecological model.

Per this theoretical construction, each system contains roles, norms and rules which may shape psychological development. For example, an inner-city family faces many challenges which an affluent family in a gated community does not, and vice versa. The inner-city family is more likely to experience environmental hardships, like crime and squalor. On the other hand, the sheltered family is more likely to lack the nurturing support of extended family.

Since its publication in 1979, Bronfenbrenner's major statement of this theory, The Ecology of Human Development has had widespread influence on the way psychologists and others approach the study of human beings and their environments. As a result of his groundbreaking work in human ecology, these environments — from the family to economic and political structures — have come to be viewed as part of the life course from childhood through adulthood.

Bronfenbrenner has identified Soviet developmental psychologist Lev Vygotsky and German-born psychologist Kurt Lewin as important influences on his theory.

Bronfenbrenner's work provides one of the foundational elements of the ecological counseling perspective, as espoused by Robert K. Conyne, Ellen Cook, and the University of Cincinnati Counseling Program.

There are many different theories related to human development. Human ecology theory emphasizes environmental factors as central to development.

Role of Technology

Children who depend on technology for play and entertainment grossly limit their creativity and imagination as well as the optimal growth of their sensory motor skills. Bombarding sedentary young bodies with chaotic sensory stimulation can result in the delay of developmental milestones. The subsequent impact on the development of foundational literary skills, has caused France to ban all "Baby TV". Violent content found in media has had such an impact on child aggression that the United States has classified media violence as a public health risk. Students entering schools struggle with discipline and attention skills required for optimal learning, which can result in major behaviour management problems for teachers in the classroom.

Demerits

Thus Technological advancements have indirectly contributed to physical, psychological and behavioral disorders thereby directly interfering with the microsystem as suggested by Bronfenbrenner's framework.

Three critical factors for the healthy, physical and psychological development of a child are movement, touch and connection with other human beings. These are essential sensory inputs for the development of a child's motor and attachment systems, deprivation of which can have devastating consequences. Children who fail to connect with the exo- and macrosystems of Bronfenbrenner's structure are likely to be deprived of friends, family and neighbours, as well as the attitudes/ideologies of the culture within.

Merits

The evidence from the Corporation for Public Broadcasting's "Ready To Learn" initiative indicates that when television shows and electronic resources have been carefully designed to incorporate what is known about effective reading instruction, they serve as positive and powerful tools for teaching and learning. The educational content is what matters, not the format in which it is presented.

The Ecological Laws of Thermodynamics

Energy is defined as the ability to do work. The behavior of energy is described by the two laws, the first law of thermodynamics or the law of conservation of energy, states that energy may be transformed from one form into another but is neither created nor destroyed. Light, for example, is a form of energy; it can be transformed into work, heat, or potential energy of food, depending on the situation, but none of it is destroyed.

The second law of thermodynamics, or the law of entropy, may be stated as, no process involving an energy transformation will spontaneously occur unless there is a degradation of energy from a concentrated form into a dispersed form. For example, heat in a hot object will spontaneously tend to become dispersed into the cooler surroundings.

The second law of thermodynamics may also be stated as, because some energy is always dispersed into unavailable heat energy, no spontaneous transformation of energy (sunlight, for example) into potential energy (protoplasm, for example) is 100 percent efficient. Entropy (from en = "in" and trope = "transformation") is a measure of the unavailable energy resulting from transformations; the term is also used as a general index of the disorder associated with energy degradation.

Organisms, ecosystems, and the entire ecosphere possess the essential thermodynamic characteristic- they can create and maintain a high state of internal order, or a condition of low entropy (a low amount of disorder). Low entropy is achieved by continually and efficiently dissipating energy of high utility (light or food, for example) into energy of low utility (heat, for example).

In the ecosystem, order in a complex biomass structure is maintained by the total community respiration, which continually "pumps out disorder". Accordingly, ecosystems and organisms are open, non-equilibrium thermodynamic systems that continuously exchange energy and matter with the environment to decrease internal entropy but increase external entropy (thus conforming to the laws of thermodynamics).

The fundamental concepts of thermodynamics outlined in the preceding paragraph are the most important of the natural laws that supply to all biological or ecological systems. So far as is known, no exceptions—and no technological innovations—can break these laws of physics. Any system of humankind or nature that does not conform to them is indeed doomed.

Various forms of life are all accompanied by energy changes, even though no energy is created or destroyed (first law of thermodynamics). The energy that reaches the surface of Earth as light is balanced by the energy that leaves the surface of Earth as invisible heat radiation.

The essence of life is the progression of such changes as growth, self-duplication, and the synthesis of complex combinations of matter. Without the energy transfers that accompany all such changes, there could be no life and no ecological systems. Humankind is just one of the remarkable natural proliferations that depend on the continuous inflow of concentrated energy.

Ecologists understand how light is related to ecological systems and how energy is transformed within the system. The relationships between producer plants and consumer animals, between predators and prey, not to mention the numbers and kinds of organisms in a given environment, are all limited and controlled by the flow of energy from concentrated to dispersed forms.

Ecologists and environmental engineers are now using natural ecosystems as models in an attempt to design more energy-efficient human-built systems to transform fossil fuel, atomic energy, and other forms of concentrated energy in industrial and technological societies.

The same basic laws that govern nonliving systems, such as automobiles or computers, also govern all types of ecosystems, such as agro-ecosystems. The difference is that living systems use part of their internally available energy for self-repair and for "pumping out" disorder; machines have to be repaired and replaced by the use of external energy. In their enthusiasm for machines and technology, some forget that a considerable amount of energy resources must be reserved at all times for reducing the entropy created by their operation.

When light is absorbed by some object, which becomes warmer as a result, the light energy has been transformed into another kind of energy heat energy. Heat energy comprises the vibrations and motions of the molecules that make up an object. The differential absorption of the rays from the Sun by land and water causes hot and cold areas, leading to the flow of air, which may drive windmills and perform work, such as the pumping of water against the force of gravity.

In this case, light energy changes into heat energy on the land surface of Earth, then into kinetic energy of moving air, which accomplishes the work of raising water. The energy is not destroyed by the lifting of the water; instead, it becomes potential energy, because the latent energy inherent in having the water elevated can then be transformed into some other type of energy by allowing the water to fall back down to its original level.

Food resulting from the photosynthesis of green plants represents potential energy, which changes into other forms of energy when the food is used by organisms. Because the amount of one type of energy is always equivalent in quantity (but not in quality) to another type into which it is transformed, we can calculate one from the other. Energy that is "consumed" is not actually used up.

Rather, it is converted from a state of high-quality to a state of low-quality energy. Gasoline in the tank of an automobile is indeed used up as gasoline; however, the energy in the tank is not destroyed but converted into forms no longer usable by the automobile.

The second law of thermodynamics deals with the transfer of energy toward an ever less available and more dispersed state. As far as the solar system is concerned, the ultimate dispersed state is one in which all energy ends up in the form of evenly distributed heat energy. This degradation process has often been spoken of as "the running down of the solar system".

At present, Earth is far from that stable state of energy, because vast potential energy and temperature differences are maintained by the continual influx of energy from the Sun. However, the process of going toward the stable state is responsible for the succession of energy changes that constitute natural phenomena on Earth. The situation is rather like that of a person on a treadmill; that person never reaches the end of the treadmill, but the effort to do so results in well-defined physiological and health-related processes.

Thus, when the energy of the Sun strikes Earth, it tends to be degraded into heat energy. Only a very small portion (less than 1 percent) of the light energy absorbed by green plants is transformed into potential or food energy; most of it goes into heat, which then passes out of the plant, the ecosystem, and the ecosphere. The rest of the biological world obtains its potential chemical energy from the organic substances produced by plant photosynthesis or microorganism chemosynthesis.

An animal, for example, takes the chemical potential energy of food and converts a large part of it into heat in order to enable a small part of the energy to be reestablished as the chemical potential energy of new protoplasm. At each step in the transfer of energy from one organism to another, a large part of the energy is degraded into heat. However, entropy is not all negative. As the quantity of available energy declines, the quality of the remainder may be greatly enhanced.

Over the years, many theorists were bothered by the fact that the functional order maintained within living systems seemed to defy the second law of thermodynamics. Ilya Prigogine, who won the Noble Prize for his work in non-equilibrium thermodynamics, resolved this apparent contradiction by showing that self-organization and the creation of new structures can and do occur in systems that are far from equilibrium and have well-developed "dissipative structures" that pump out the disorder. The respiration of the highly ordered biomass is the "dissipative structure" in an ecosystem.

Although entropy in the technical sense relates to energy, the word is also used in a broader sense to refer to the degradation of matter. Freshly made steel represents a low-entropy (high-utility) state of iron; the rusting frame of an automobile represents a high-entropy (low-utility) state. Accordingly, a high-entropy civilization is characterized by degrading energy, such as dilapidating infrastructures (rusting pipes, rotting wood) or eroding soil. Constant repair is one of the costs of high- energy civilizations.

There are two classes of basic units: potential energy units, independent of time (Class A), and power or rate units, with time built into the definition (Class B). Inter-conversions of power units must take account of the time unit used; thus, 1 watt = 860 cal/h. Of course, Class A units become power units if a time period is included (for example, BTU per hour, day, or year), and power units can be converted back to energy units by "multiplying out" the time unit (as in the case of KWh).

The transfer of energy through the food chain of an ecosystem is termed the energy flow because, in accordance with the law of entropy, energy transformations are "one way" in contrast to the cyclic behavior of matter.

Energy Flow in Ecosystem

Energy moves life. The cycle of energy is based on the flow of energy through different trophic levels in an ecosystem. Our ecosystem is maintained by the cycling energy and nutrients obtained from different external sources. At the first trophic level, primary producers use solar energy to produce organic material through photosynthesis.

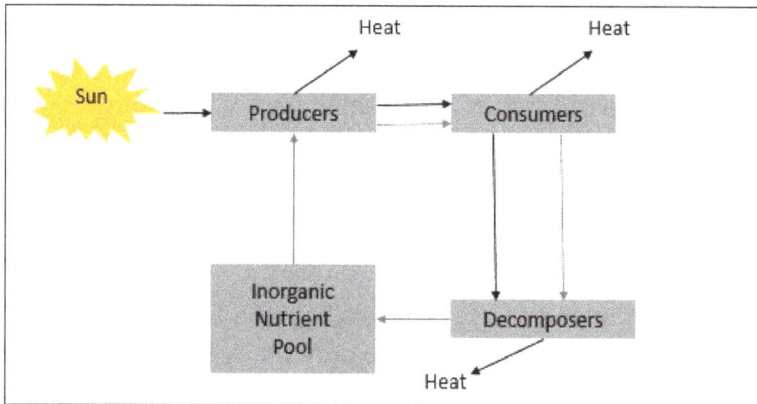

The herbivores at the second trophic level, use the plants as food which gives them energy. A large part of this energy is used up for the metabolic functions of these animals such as breathing, digesting food, supporting growth of tissues, maintaining blood circulation and body temperature.

The carnivores at the next trophic level, feed on the herbivores and derive energy for their sustenance and growth. If large predators are present, they represent still higher trophic level and they feed on carnivores to get energy. Thus, the different plants and animal species are linked to one another through food chains.

Decomposers which include bacteria, fungi, molds, worms, and insects break down wastes and dead organisms, and return the nutrients to the soil, which is then taken up by the producers. Energy is not recycled during decomposition, but it is released.

Biogeochemical Cycles

All elements in the earth are recycled time and again. The major elements such as oxygen, carbon, nitrogen, phosphorous, and sulphur are essential ingredients that make up organisms.

Biogeochemical cycles refer to the flow of such chemical elements and compounds between organisms and the physical environment. Chemicals taken in by organisms are passed through the food chain and come back to the soil, air, and water through mechanisms such as respiration, excretion, and decomposition.

As an element moves through this cycle, it often forms compounds with other elements as a result of metabolic processes in living tissues and of natural reactions in the atmosphere, hydrosphere, or lithosphere.

Such cyclic exchange of material between the living organisms and their non-living environment is called Biogeochemical Cycle.

Following are some important biogeochemical cycles:

- Carbon Cycle,

- Nitrogen Cycle,

- Water Cycle,

- Oxygen Cycle,

- Phosphorus Cycle,

- Sulphur Cycle.

Carbon Cycle

Carbon enters into the living world in the form of carbon dioxide through the process of photosynthesis as carbohydrates. These organic compounds (food) are then passed from the producers to the consumers (herbivores & carnivores). This carbon is finally returned to the surrounding medium by the process of respiration or decomposition of plants and animals by the decomposers. Carbon is also recycled during the burning of fossil fuels.

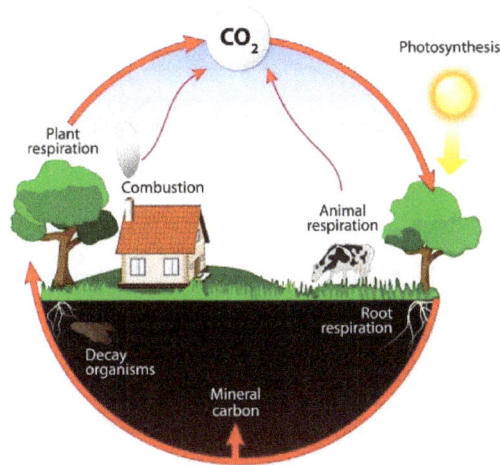

Nitrogen Cycle

Nitrogen is present in the atmosphere in an elemental form and as such it cannot be utilized by living organisms. This elemental form of nitrogen is converted into combined state with elements such as H, C, O by certain bacteria, so that it can be readily used by the plants.

Nitrogen is being continuously expelled into the air by the action of microorganisms such as denitrifying bacteria and finally returned to the cycle through the action of lightening and electrification.

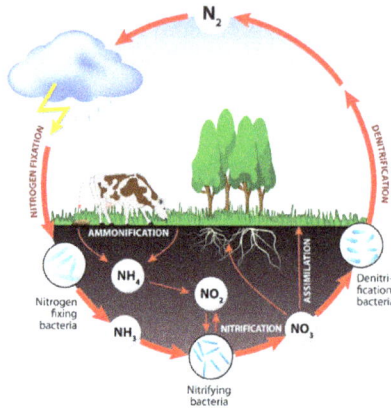

Water Cycle

The evaporation of water from ocean, rivers, lakes, and transpiring plants takes water in the form of vapors to the atmosphere. This vaporized water subsequently cools and condenses to form cloud and water. This cooled water vapor ultimately returns to the earth as rain and snow, completing the cycle.

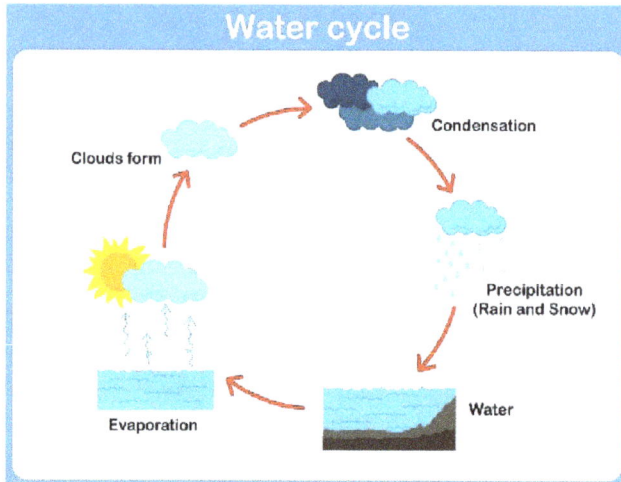

Nutrient Cycle

A nutrient cycle (or ecological recycling) is the movement and exchange of organic and inorganic matter back into the production of matter. Energy flow is a unidirectional and noncyclic pathway, whereas the movement of mineral nutrients is cyclic. Mineral cycles include the carbon cycle, sulfur cycle, nitrogen cycle, water cycle, phosphorus cycle, oxygen cycle, among others that continually recycle along with other mineral nutrients into productive ecological nutrition.

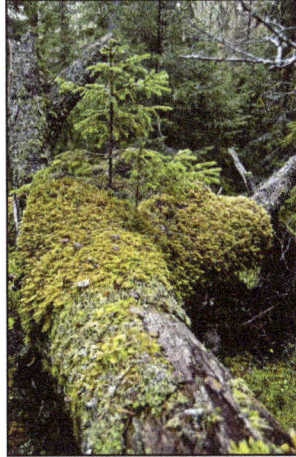

Fallen logs are critical components of the nutrient cycle in terrestrial forests.
Nurse logs form habitats for other creatures that decompose the materials
and recycle the nutrients back into production.

The nutrient cycle is nature's recycling system. All forms of recycling have feedback loops that use energy in the process of putting material resources back into use. Recycling in ecology is regulated to a large extent during the process of decomposition. Ecosystems employ biodiversity in the food webs that recycle natural materials, such as mineral nutrients, which includes water. Recycling in natural systems is one of the many ecosystem services that sustain and contribute to the well-being of human societies.

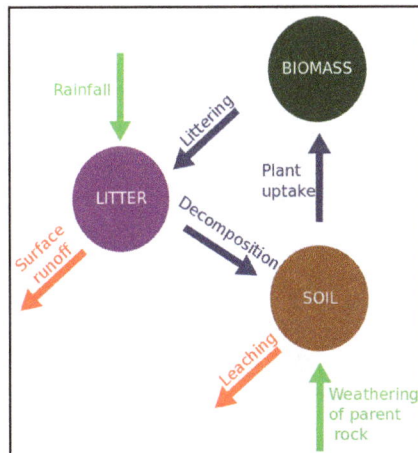

A nutrient cycle of a typical terrestrial ecosystem.

There is much overlap between the terms for the biogeochemical cycle and nutrient cycle. Most textbooks integrate the two and seem to treat them as synonymous terms. However, the terms often appear independently. Nutrient cycle is more often used in direct reference to the idea of an intra-system cycle, where an ecosystem functions as a unit. From a practical point, it does not make sense to assess a terrestrial ecosystem by considering the full column of air above it as well as the great depths of Earth below it. While an ecosystem often has no clear boundary, as a working model it is practical to consider the functional community where the bulk of matter and energy transfer occurs. Nutrient cycling occurs in ecosystems that participate in the "larger biogeochemical cycles of the earth through a system of inputs and outputs".

Complete and Closed Loop

Ecosystems are capable of complete recycling. Complete recycling means that 100% of the waste material can be reconstituted indefinitely. This idea was captured by Howard T. Odum when he penned that "it is thoroughly demonstrated by ecological systems and geological systems that all the chemical elements and many organic substances can be accumulated by living systems from background crustal or oceanic concentrations without limit as to concentration so long as there is available solar or another source of potential energy" In 1979 Nicholas Georgescu-Roegen proposed the fourth law of entropy stating that complete recycling is impossible. Despite Georgescu-Roegen's extensive intellectual contributions to the science of ecological economics, the fourth law has been rejected in line with observations of ecological recycling. However, some authors state that complete recycling is impossible for technological waste.

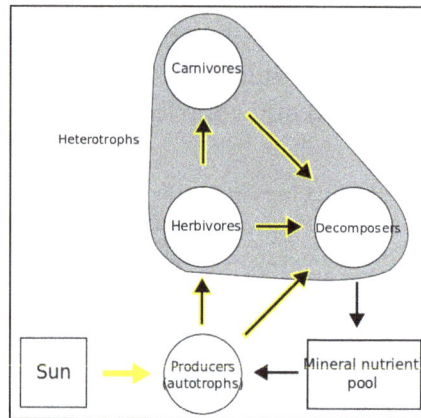

A simplified food web illustrating a three-trophic food chain (producers-herbivores-carnivores) linked to decomposers. The movement of mineral nutrients through the food chain, into the mineral nutrient pool, and back into the trophic system illustrates ecological recycling. The movement of energy, in contrast, is unidirectional and noncyclic.

Ecosystems execute closed loop recycling where demand for the nutrients that adds to the growth of biomass exceeds supply within that system. There are regional and spatial differences in the rates of growth and exchange of materials, where some ecosystems may be in nutrient debt (sinks) where others will have extra supply (sources). These differences relate to climate, topography, and geological history leaving behind different sources of parent material. In terms of a food web, a cycle or loop is defined as "a directed sequence of one or more links starting from, and ending at, the same species". An example of this is the microbial food web in the ocean, where "bacteria are exploited, and controlled, by protozoa, including heterotrophic microflagellates which are in turn exploited by ciliates. This grazing activity is accompanied by excretion of substances which are in turn used by the bacteria so that the system more or less operates in a closed circuit".

Ecological Recycling

An example of ecological recycling occurs in the enzymatic digestion of cellulose. "Cellulose, one of the most abundant organic compounds on Earth, is the major polysaccharide in plants where it is part of the cell walls. Cellulose-degrading enzymes participate in the natural, ecological recycling of plant material". Different ecosystems can vary in their recycling rates of litter, which creates a complex feedback on factors such as the competitive dominance of certain plant species. Different rates and patterns of ecological recycling leaves a legacy of environmental effects with implications for the future evolution of ecosystems.

Ecological recycling is common in organic farming, where nutrient management is fundamentally different compared to agri-business styles of soil management. Organic farms that employ ecosystem recycling to a greater extent support more species (increased levels of biodiversity) and have a different food web structure. Organic agricultural ecosystems rely on the services of biodiversity for the recycling of nutrients through soils instead of relying on the supplementation of synthetic fertilizers. The model for ecological recycling agriculture adheres to the following principals:

- Protection of biodiversity.
- Use of renewable energy.
- Recycling of plant nutrients.

Where produce from an organic farm leaves the farm gate for the market the system becomes an open cycle and nutrients may need to be replaced through alternative methods.

Ecosystem Engineers

An illustration of an earthworm casting taken from Charles Darwin's publication on the movement of organic matter in soils through the ecological activities of worms.

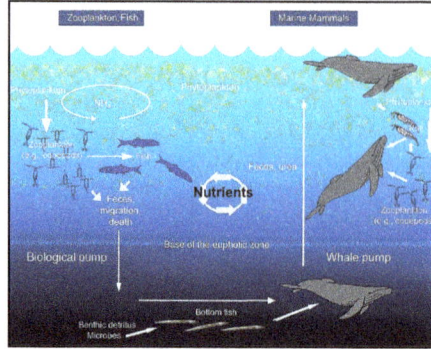

From the largest to the smallest of creatures, nutrients are recycled by their movement, by their wastes, and by their metabolic activities. This illustration shows an example of the whale pump that cycles nutrients through the layers of the oceanic water column. Whales can migrate to great depths to feed on bottom fish (such as sand lance Ammodytes spp.) and surface to feed on krill and plankton at shallower levels. The whale pump enhances growth and productivity in other parts of the ecosystem.

The persistent legacy of environmental feedback that is left behind by or as an extension of the ecological actions of organisms is known as niche construction or ecosystem engineering. Many species leave an effect even after their death, such as coral skeletons or the extensive habitat modifications to a wetland by a beaver, whose components are recycled and re-used by descendants and other species living under a different selective regime through the feedback and agency of these legacy effects. Ecosystem engineers can influence nutrient cycling efficiency rates through their actions.

Earthworms, for example, passively and mechanically alter the nature of soil environments. Bodies of dead worms passively contribute mineral nutrients to the soil. The worms also mechanically modify the physical structure of the soil as they crawl about (bioturbation), digest on the molds of organic matter they pull from the soil litter. These activities transport nutrients into the mineral layers of soil. Worms discard wastes that create worm castings containing undigested materials where bacteria and other decomposers gain access to the nutrients. The earthworm is employed in this process and the production of the ecosystem depends on their capability to create feedback loops in the recycling process.

Shellfish are also ecosystem engineers because they: 1) Filter suspended particles from the water column; 2) Remove excess nutrients from coastal bays through denitrification; 3) Serve as natural coastal buffers, absorbing wave energy and reducing erosion from boat wakes, sea level rise and storms; 4) Provide nursery habitat for fish that are valuable to coastal economies.

Fungi contribute to nutrient cycling and nutritionally rearrange patches of ecosystem creating niches for other organisms. In that way fungi in growing dead wood allow xylophages to grow and develop and xylophages, in turn, affect dead wood, contributing to wood decomposition and nutrient cycling in the forest floor.

Recycling in Novel Ecosystems

An endless stream of technological waste accumulates in different spatial configurations across the planet and turns into a predator in our soils, our streams, and our oceans. This idea was similarly expressed in 1954 by ecologist Paul Sears: "We do not know whether to cherish the forest as a source of essential raw materials and other benefits or to remove it for the space it occupies. We expect a river to serve as both vein and artery carrying away waste but bringing usable material in the same channel. Nature long ago discarded the nonsense of carrying poisonous wastes and nutrients in the same vessels". Ecologists use population ecology to model contaminants as competitors or predators. Rachel Carson was an ecological pioneer in this area as her book Silent Spring inspired research into biomagnification and brought to the world's attention the unseen pollutants moving into the food chains of the planet.

In contrast to the planets natural ecosystems, technology (or technoecosystems) is not reducing its impact on planetary resources. Only 7% of total plastic waste (adding up to millions upon millions of tons) is being recycled by industrial systems; the 93% that never makes it into the industrial recycling stream is presumably absorbed by natural recycling systems In contrast and over extensive lengths of time (billions of years) ecosystems have maintained a consistent balance with production roughly equaling respiratory consumption rates. The balanced recycling efficiency of nature means that production of decaying waste material has exceeded rates of recyclable consumption into food chains equal to the global stocks of fossilized fuels that escaped the chain of decomposition.

Microplastics and nanosilver materials flowing and cycling through ecosystems from pollution and discarded technology are among a growing list of emerging ecological concerns. For example, unique assemblages of marine microbes have been found to digest plastic accumulating in the worlds oceans. Discarded technology is absorbed into soils and creates a new class of soils called technosols. Human wastes in the Anthropocene are creating new systems of ecological recycling, novel ecosystems that have to contend with the mercury cycle and other synthetic materials that are streaming into the biodegradation chain. Microorganisms have a significant role in the removal of synthetic organic compounds from the environment empowered by recycling mechanisms that have complex biodegradation pathways. The effect of synthetic materials, such as nanoparticles and microplastics, on ecological recycling systems is listed as one of the major concerns for ecosystem in this century.

Technological Recycling

Recycling in human industrial systems (or technoecosystems) differs from ecological recycling in scale, complexity, and organization. Industrial recycling systems do not focus on the employment of ecological food webs to recycle waste back into different kinds of marketable goods, but primarily employ people and technodiversity instead. Some researchers have questioned the premise behind these and other kinds of technological solutions under the banner of 'eco-efficiency' are limited in their capability, harmful to ecological processes, and dangerous in their hyped capabilities. Many technoecosystems are competitive and parasitic toward natural ecosystems. Food web or biologically based "recycling includes metabolic recycling (nutrient recovery, storage, etc.) and ecosystem recycling (leaching and in situ organic matter mineralization, either in the water column, in the sediment surface, or within the sediment)".

Food Chain

A food chain is a linear network of links in a food web starting from producer organisms (such as grass or trees which use radiation from the Sun to make their food) and ending at apex predator species (like grizzly bears or killer whales), detritivores (like earthworms or woodlice), or decomposer species (such as fungi or bacteria). A food chain also shows how the organisms are related with each other by the food they eat. Each level of a food chain represents a different trophic level. A food chain differs from a food web, because the complex network of different animals' feeding relations are aggregated and the chain only follows a direct, linear pathway of one animal at a time. Natural interconnections between food chains make it a food web. A common metric used to the quantify food web trophic structure is food chain length. In its simplest form, the length of a chain is the number of links between a trophic consumer and the base of the web and the mean chain length of an entire web is the arithmetic average of the lengths of all chains in a food web.

Many food webs have a keystone species (such as sharks). A keystone species is a species that has a large impact on the surrounding environment and can directly affect the food chain. If this keystone species dies off it can set the entire food chain off balance. Keystone species keep herbivores from depleting all of the foliage in their environment and preventing a mass extinction.

Food Chain Length

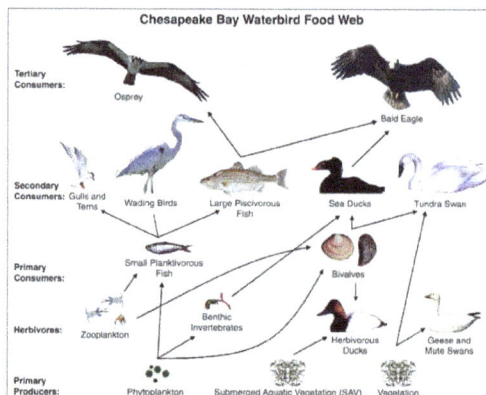

This food web of waterbirds from Chesapeake Bay is a network of food chains.

The food chain's length is a continuous variable that provides a measure of the passage of energy and an index of ecological structure that increases in value counting progressively through the linkages in a linear fashion from the lowest to the highest trophic (feeding) levels.

Food chains are often used in ecological modeling (such as a three species food chain). They are simplified abstractions of real food webs, but complex in their dynamics and mathematical implications.

Ecologists have formulated and tested hypotheses regarding the nature of ecological patterns associated with food chain length, such as increasing length increasing with ecosystem size, reduction of energy at each successive level, or the proposition that long food chain lengths are unstable. Food chain studies have an important role in ecotoxicology studies tracing the pathways and biomagnification of environmental contaminants.

Producers, such as plants, are organisms that utilize solar or chemical energy to synthesize starch. All food chains must start with a producer. In the deep sea, food chains centered on hydrothermal vents and cold seeps exist in the absence of sunlight. Chemosynthetic bacteria and archaea use hydrogen sulfide and methane from hydrothermal vents and cold seeps as an energy source (just as plants use sunlight) to produce carbohydrates; they form the base of the food chain. Consumers are organisms that eat other organisms. All organisms in a food chain, except the first organism, are consumers.

In a food chain, there is also reliable energy transfer through each stage. However, all the energy at one stage of the chain is not absorbed by the organism at the next stage. The amount of energy from one stage to another decreases.

Food Web

A food web (or food cycle) is the natural interconnection of food chains and a graphical representation (usually an image) of what-eats-what in an ecological community. Another name for food web is consumer-resource system. Ecologists can broadly lump all life forms into one of two categories called trophic levels: 1) the autotrophs, and 2) the heterotrophs. To maintain their bodies, grow, develop, and to reproduce, autotrophs produce organic matter from inorganic substances, including both minerals and gases such as carbon dioxide. These chemical reactions require energy, which mainly comes from the Sun and largely by photosynthesis, although a very small amount comes from bioelectrogenesis in wetlands, and mineral electron donors in hydrothermal vents and hot springs. A gradient exists between trophic levels running from complete autotrophs that obtain their sole source of carbon from the atmosphere, to mixotrophs (such as carnivorous plants) that are autotrophic organisms that partially obtain organic matter from sources other than the atmosphere, and complete heterotrophs that must feed to obtain organic matter. The linkages in a food web illustrate the feeding pathways, such as where heterotrophs obtain organic matter by feeding on autotrophs and other heterotrophs. The food web is a simplified illustration of the various methods of feeding that links an ecosystem into a unified system of exchange. There are different kinds of feeding relations that can be roughly divided into herbivory, carnivory, scavenging and parasitism. Some of the organic matter eaten by heterotrophs, such as sugars, provides energy. Autotrophs and heterotrophs come in all sizes, from microscopic to many tonnes - from cyanobacteria to giant redwoods, and from viruses and bdellovibrio to blue whales.

Charles Elton pioneered the concept of food cycles, food chains, and food size in his classical 1927 book "Animal Ecology"; Elton's 'food cycle' was replaced by 'food web' in a subsequent ecological text. Elton organized species into functional groups, which was the basis for Raymond Lindeman's classic and landmark paper in 1942 on trophic dynamics. Lindeman emphasized the important role of decomposer organisms in a trophic system of classification. The notion of a food web has a historical foothold in the writings of Charles Darwin and his terminology, including an "entangled bank", "web of life", "web of complex relations", and in reference to the decomposition actions of earthworms he talked about "the continued movement of the particles of earth". Even earlier, in 1768 John Bruckner described nature as "one continued web of life".

Food webs are limited representations of real ecosystems as they necessarily aggregate many species into trophic species, which are functional groups of species that have the same predators and prey in a food web. Ecologists use these simplifications in quantitative (or mathematical representation) models of trophic or consumer-resource systems dynamics. Using these models they can measure and test for generalized patterns in the structure of real food web networks. Ecologists have identified non-random properties in the topographic structure of food webs. Published examples that are used in meta analysis are of variable quality with omissions. However, the number of empirical studies on community webs is on the rise and the mathematical treatment of food webs using network theory had identified patterns that are common to all. Scaling laws, for example, predict a relationship between the topology of food web predator-prey linkages and levels of species richness.

Taxonomy of a Food Web

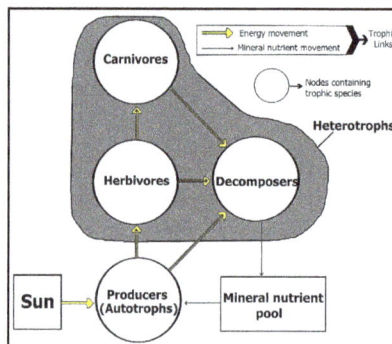

A simplified food web illustrating a three trophic food chain (producers-herbivores-carnivores) linked to decomposers. The movement of mineral nutrients is cyclic, whereas the movement of energy is unidirectional and noncyclic. Trophic species are encircled as nodes and arrows depict the links.

Links in food webs map the feeding connections (who eats whom) in an ecological community. Food cycle is an obsolete term that is synonymous with food web. Ecologists can broadly group all life forms into one of two trophic layers, the autotrophs and the heterotrophs. Autotrophs produce more biomass energy, either chemically without the sun's energy or by capturing the sun's energy in photosynthesis, than they use during metabolic respiration. Heterotrophs consume rather than produce biomass energy as they metabolize, grow, and add to levels of secondary production. A food web depicts a collection of polyphagous heterotrophic consumers that network and cycle the flow of energy and nutrients from a productive base of self-feeding autotrophs.

The base or basal species in a food web are those species without prey and can include autotrophs or saprophytic detritivores (i.e., the community of decomposers in soil, biofilms, and periphyton). Feeding connections in the web are called trophic links. The number of trophic links per consumer is a measure of food web connectance. Food chains are nested within the trophic links of food webs. Food chains are linear (noncyclic) feeding pathways that trace monophagous consumers from a base species up to the top consumer, which is usually a larger predatory carnivore.

Linkages connect to nodes in a food web, which are aggregates of biological taxa called trophic species. Trophic species are functional groups that have the same predators and prey in a food web. Common examples of an aggregated node in a food web might include parasites, microbes, decomposers, saprotrophs, consumers, or predators, each containing many species in a web that can otherwise be connected to other trophic species.

Trophic Levels

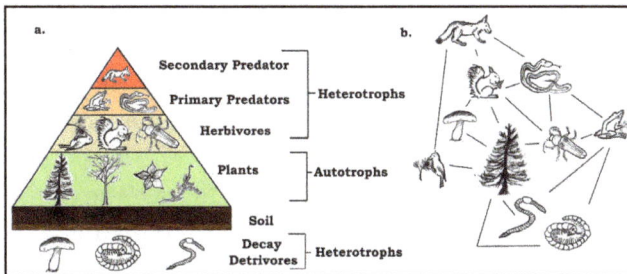

A trophic pyramid (a) and a simplified community food web (b) illustrating ecological relations among creatures that are typical of a northern Boreal terrestrial ecosystem. The trophic pyramid roughly represents the biomass (usually measured as total dry-weight) at each level. Plants generally have the greatest biomass. Names of trophic categories are shown to the right of the pyramid. Some ecosystems, such as many wetlands, do not organize as a strict pyramid, because aquatic plants are not as productive as long-lived terrestrial plants such as trees. Ecological trophic pyramids are typically one of three kinds: 1) pyramid of numbers, 2) pyramid of biomass, or 3) pyramid of energy.

Food webs have trophic levels and positions. Basal species, such as plants, form the first level and are the resource limited species that feed on no other living creature in the web. Basal species can be autotrophs or detritivores, including "decomposing organic material and its associated microorganisms which we defined as detritus, micro-inorganic material and associated microorganisms (MIP), and vascular plant material". Most autotrophs capture the sun's energy in chlorophyll, but some autotrophs (the chemolithotrophs) obtain energy by the chemical oxidation of inorganic compounds and can grow in dark environments, such as the sulfur bacterium Thiobacillus, which lives in hot sulfur springs. The top level has top (or apex) predators which no other species kills directly for its food resource needs. The intermediate levels are filled with omnivores that feed on more than one trophic level and cause energy to flow through a number of food pathways starting from a basal species.

In the simplest scheme, the first trophic level (level 1) is plants, then herbivores (level 2), and then carnivores (level 3). The trophic level is equal to one more than the chain length, which is the number of links connecting to the base. The base of the food chain (primary producers or detritivores) is set at zero. Ecologists identify feeding relations and organize species into trophic species through extensive gut content analysis of different species. The technique has been improved through the use of stable isotopes to better trace energy flow through the web. It was once thought that omnivory was rare, but recent evidence suggests otherwise. This realization has made trophic classifications more complex.

Trophic Dynamics

The trophic level concept was introduced in a historical landmark paper on trophic dynamics in 1942 by Raymond L. Lindeman. The basis of trophic dynamics is the transfer of energy from one part of the ecosystem to another. The trophic dynamic concept has served as a useful quantitative heuristic, but it has several major limitations including the precision by which an organism can be allocated to a specific trophic level. Omnivores, for example, are not restricted to any single level. Nonetheless, recent research has found that discrete trophic levels do exist, but "above the herbivore trophic level, food webs are better characterized as a tangled web of omnivores".

A central question in the trophic dynamic literature is the nature of control and regulation over resources and production. Ecologists use simplified one trophic position food chain models (producer, carnivore, decomposer). Using these models, ecologists have tested various types of ecological control mechanisms. For example, herbivores generally have an abundance of vegetative resources, which meant that their populations were largely controlled or regulated by predators. This is known as the top-down hypothesis or 'green-world' hypothesis. Alternatively to the top-down hypothesis, not all plant material is edible and the nutritional quality or antiherbivore defenses of plants (structural and chemical) suggests a bottom-up form of regulation or control. Recent studies have concluded that both "top-down" and "bottom-up" forces can influence

community structure and the strength of the influence is environmentally context dependent. These complex multitrophic interactions involve more than two trophic levels
in a food web.

Another example of a multi-trophic interaction is a trophic cascade, in which predators help to increase plant growth and prevent overgrazing by suppressing herbivores.
Links in a food-web illustrate direct trophic relations among species, but there are also
indirect effects that can alter the abundance, distribution, or biomass in the trophic levels. For example, predators eating herbivores indirectly influence the control and regulation of primary production in plants. Although the predators do not eat the plants
directly, they regulate the population of herbivores that are directly linked to plant tro-
phism. The net effect of direct and indirect relations is called trophic cascades. Trophic
cascades are separated into species-level cascades, where only a subset of the food-web
dynamic is impacted by a change in population numbers, and community-level cascades, where a change in population numbers has a dramatic effect on the entire food-
web, such as the distribution of plant biomass.

Energy Flow and Biomass

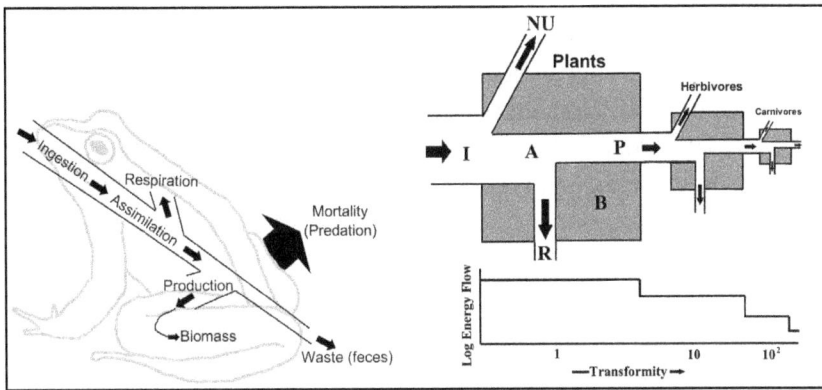

Left: Energy flow diagram of a frog. The frog represents a node in an extended food web.
The energy ingested is utilized for metabolic processes and transformed into biomass.
The energy flow continues on its path if the frog is ingested by predators, parasites, or
as a decaying carcass in soil. This energy flow diagram illustrates how energy is lost as
it fuels the metabolic process that transform the energy and nutrients into biomass.

Right: An expanded three link energy food chain (1. plants, 2. herbivores, 3. carnivores)
illustrating the relationship between food flow diagrams and energy transformity. The
transformity of energy becomes degraded, dispersed, and diminished from higher quality to lesser quantity as the energy within a food chain flows from one trophic species
into another. Abbreviations: I=input, A=assimilation, R=respiration, NU=not utilized,
P=production, B=biomass.

Food webs depict energy flow via trophic linkages. Energy flow is directional, which
contrasts against the cyclic flows of material through the food web systems. Energy

flow "typically includes production, consumption, assimilation, non-assimilation losses (feces), and respiration (maintenance costs)". In a very general sense, energy flow (E) can be defined as the sum of metabolic production (P) and respiration (R), such that E=P+R.

Biomass represents stored energy. However, concentration and quality of nutrients and energy is variable. Many plant fibers, for example, are indigestible to many herbivores leaving grazer community food webs more nutrient limited than detrital food webs where bacteria are able to access and release the nutrient and energy stores. "Organisms usually extract energy in the form of carbohydrates, lipids, and proteins. These polymers have a dual role as supplies of energy as well as building blocks; the part that functions as energy supply results in the production of nutrients (and carbon dioxide, water, and heat). Excretion of nutrients is, therefore, basic to metabolism". The units in energy flow webs are typically a measure mass or energy per m2 per unit time. Different consumers are going to have different metabolic assimilation efficiencies in their diets. Each trophic level transforms energy into biomass. Energy flow diagrams illustrate the rates and efficiency of transfer from one trophic level into another and up through the hierarchy.

It is the case that the biomass of each trophic level decreases from the base of the chain to the top. This is because energy is lost to the environment with each transfer as entropy increases. About eighty to ninety percent of the energy is expended for the organism's life processes or is lost as heat or waste. Only about ten to twenty percent of the organism's energy is generally passed to the next organism. The amount can be less than one percent in animals consuming less digestible plants, and it can be as high as forty percent in zooplankton consuming phytoplankton. Graphic representations of the biomass or productivity at each tropic level are called ecological pyramids or trophic pyramids. The transfer of energy from primary producers to top consumers can also be characterized by energy flow diagrams.

Material Flux and Recycling

Many of the Earth's elements and minerals (or mineral nutrients) are contained within the tissues and diets of organisms. Hence, mineral and nutrient cycles trace food web energy pathways. Ecologists employ stoichiometry to analyze the ratios of the main elements found in all organisms: carbon (C), nitrogen (N), phosphorus (P). There is a large transitional difference between many terrestrial and aquatic systems as C:P and C:N ratios are much higher in terrestrial systems while N:P ratios are equal between the two systems. Mineral nutrients are the material resources that organisms need for growth, development, and vitality. Food webs depict the pathways of mineral nutrient cycling as they flow through organisms. Most of the primary production in an ecosystem is not consumed, but is recycled by detritus back into useful nutrients. Many of the Earth's microorganisms are involved in the formation of minerals in a process called biomineralization. Bacteria that live in detrital sediments create and cycle nutrients and biominerals. Food web models and nutrient cycles have traditionally been treated separately, but there is a strong functional connection between the two in terms of stability, flux, sources, sinks, and recycling of mineral nutrients.

Kinds of Food Webs

Food webs are necessarily aggregated and only illustrate a tiny portion of the complexity of real ecosystems. For example, the number of species on the planet are likely in the general order of 107, over 95% of these species consist of microbes and invertebrates, and relatively few have been named or classified by taxonomists. It is explicitly understood that natural systems are 'sloppy' and that food web trophic positions simplify the complexity of real systems that sometimes overemphasize many rare interactions. Most studies focus on the larger influences where the bulk of energy transfer occurs. "These omissions and problems are causes for concern, but on present evidence do not present insurmountable difficulties".

There are different kinds or categories of food webs:

- Source web - One or more node(s), all of their predators, all the food these predators eat, and so on.

- Sink web - One or more node(s), all of their prey, all the food that these prey eat, and so on.

- Community (or connectedness) web - A group of nodes and all the connections of who eats whom.

- Energy flow web - Quantified fluxes of energy between nodes along links between a resource and a consumer.

- Paleoecological web - A web that reconstructs ecosystems from the fossil record.

- Functional web - Emphasizes the functional significance of certain connections having strong interaction strength and greater bearing on community organization, more so than energy flow pathways. Functional webs have compartments, which are sub-groups in the larger network where there are different densities and strengths of interaction. Functional webs emphasize that "the importance of each population in maintaining the integrity of a community is reflected in its influence on the growth rates of other populations".

Within these categories, food webs can be further organized according to the different kinds of ecosystems being investigated. For example, human food webs, agricultural food webs, detrital food webs, marine food webs, aquatic food webs, soil food webs, Arctic (or polar) food webs, terrestrial food webs, and microbial food webs. These characterizations stem from the ecosystem concept, which assumes that the phenomena under investigation (interactions and feedback loops) are sufficient to explain patterns within boundaries, such as the edge of a forest, an island, a shoreline, or some other pronounced physical characteristic.

Detrital Web

In a detrital web, plant and animal matter is broken down by decomposers, e.g., bacteria and fungi, and moves to detritivores and then carnivores. There are often relationships between the detrital web and the grazing web. Mushrooms produced by decomposers in the detrital web become a food source for deer, squirrels, and mice in the grazing web. Earthworms eaten by robins are detritivores consuming decaying leaves.

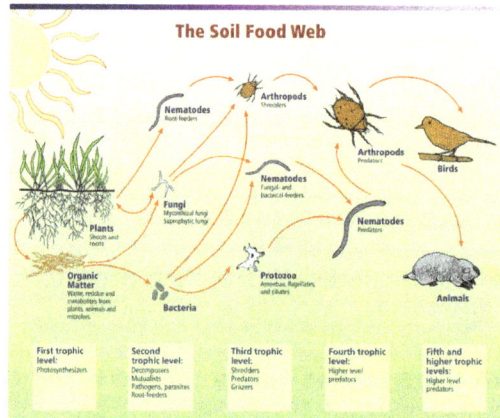

An illustration of a soil food web.

"Detritus can be broadly defined as any form of non-living organic matter, including different types of plant tissue (e.g. leaf litter, dead wood, aquatic macrophytes, algae), animal tissue (carrion), dead microbes, faeces (manure, dung, faecal pellets, guano, frass), as well as products secreted, excreted or exuded from organisms (e.g. extra-cellular polymers, nectar, root exudates and leachates, dissolved organic matter, extra-cellular matrix, mucilage). The relative importance of these forms of detritus, in terms of origin, size and chemical composition, varies across ecosystems".

Quantitative Food Webs

Ecologists collect data on trophic levels and food webs to statistically model and mathematically calculate parameters, such as those used in other kinds of network analysis (e.g., graph theory), to study emergent patterns and properties shared among ecosystems. There are different ecological dimensions that can be mapped to create more complicated food webs, including: species composition (type of species), richness (number of species), biomass (the dry weight of plants and animals), productivity (rates of conversion of energy and nutrients into growth), and stability (food webs over time). A food web diagram illustrating species composition shows how change in a single species can directly and indirectly influence many others. Microcosm studies are used to simplify food web research into semi-isolated units such as small springs, decaying logs, and laboratory experiments using organisms that reproduce quickly, such as daphnia feeding on algae grown under controlled environments in jars of water.

While the complexity of real food webs connections are difficult to decipher, ecologists have found mathematical models on networks an invaluable tool for gaining insight into the structure, stability, and laws of food web behaviours relative to observable outcomes. "Food web theory centers around the idea of connectance". Quantitative formulas simplify the complexity of food web structure. The number of trophic links (tL), for example, is converted into a connectance value:

$$C = \frac{t_L}{S(S-1)/2}$$

where, S(S-1)/2 is the maximum number of binary connections among S species. "Connectance (C) is the fraction of all possible links that are realized (L/S2) and represents a standard measure of food web complexity". The distance (d) between every species pair in a web is averaged to compute the mean distance between all nodes in a web (D) and multiplied by the total number of links (L) to obtain link-density (LD), which is influenced by scale dependent variables such as species richness. These formulas are the basis for comparing and investigating the nature of non-random patterns in the structure of food web networks among many different types of ecosystems.

Scaling laws, complexity, chaos, and pattern correlates are common features attributed to food web structure.

Complexity and Stability

Food webs are complex. Complexity is a measure of an increasing number of permutations and it is also a metaphorical term that conveys the mental intractability or limits concerning unlimited algorithmic possibilities. In food web terminology, complexity is a product of the number of species and connectance. Connectance is "the fraction of all possible links that are realized in a network". These concepts were derived and stimulated through the suggestion that complexity leads to stability in food webs, such as increasing the number of trophic levels in more species rich ecosystems. This hypothesis was challenged through mathematical models suggesting otherwise, but subsequent studies have shown that the premise holds in real systems.

At different levels in the hierarchy of life, such as the stability of a food web, "the same overall structure is maintained in spite of an ongoing flow and change of components". The farther a living system (e.g., ecosystem) sways from equilibrium, the greater its complexity. Complexity has multiple meanings in the life sciences and in the public sphere that confuse its application as a precise term for analytical purposes in science. Complexity in the life sciences (or biocomplexity) is defined by the "properties emerging from the interplay of behavioral, biological, physical, and social interactions that affect, sustain, or are modified by living organisms, including humans".

Several concepts have emerged from the study of complexity in food webs. Complexity explains many principals pertaining to self-organization, non-linearity, interaction, cybernetic feedback, discontinuity, emergence, and stability in food webs. Nestedness, for example, is defined as "a pattern of interaction in which specialists interact with species that form perfect subsets of the species with which generalists interact", "—that is, the diet of the most specialized species is a subset of the diet of the next more generalized species, and its diet a subset of the next more generalized, and so on". Until recently, it was thought that food webs had little nested structure, but empirical evidence shows that many published webs have nested subwebs in their assembly.

Food webs are complex networks. As networks, they exhibit similar structural properties and mathematical laws that have been used to describe other complex systems, such as small world and scale free properties. The small world attribute refers to the many loosely connected nodes, non-random dense clustering of a few nodes (i.e., trophic or keystone species in ecology), and small path length compared to a regular lattice. "Ecological networks, especially mutualistic networks, are generally very heterogeneous, consisting of areas with sparse links among species and distinct areas of tightly linked species. These regions of high link density are often referred to as cliques, hubs, compartments, cohesive sub-groups, or modules. Within food webs, especially in aquatic systems, nestedness appears to be related to body size because the diets of smaller predators tend to be nested subsets of those of larger predators, and phylogenetic constraints, whereby related taxa are nested based on their common evolutionary history, are also evident. "Compartments in food webs are subgroups of taxa in which many strong interactions occur within the subgroups and few weak interactions occur between the subgroups. Theoretically, compartments increase the stability in networks, such as food webs".

Food webs are also complex in the way that they change in scale, seasonally, and geographically. The components of food webs, including organisms and mineral nutrients, cross the thresholds of ecosystem boundaries. This has led to the concept or area of study known as cross-boundary subsidy. "This leads to anomalies, such as food web calculations determining that an ecosystem can support one half of a top carnivore, without specifying which end". Nonetheless, real differences in structure and function have been identified when comparing different kinds of ecological food webs, such as terrestrial vs. aquatic food webs.

Ecological Pyramid

An ecological pyramid (also trophic pyramid, Eltonian pyramid, energy pyramid, or sometimes food pyramid) is a graphical representation designed to show the biomass or bioproductivity at each trophic level in a given ecosystem.

A pyramid of energy shows how much energy is retained in the form of new biomass at each trophic level, while a pyramid of biomass shows how much biomass (the amount of living or organic matter present in an organism) is present in the organisms. There is also a pyramid of numbers representing the number of individual organisms at each trophic level. Pyramids of energy are normally upright, but other pyramids can be inverted or take other shapes.

Ecological pyramids begin with producers on the bottom (such as plants) and proceed through the various trophic levels (such as herbivores that eat plants, then carnivores that eat flesh, then omnivores that eat both plants and flesh, and so on). The highest level is the top of the food chain.

Biomass can be measured by a bomb calorimeter.

Pyramid of Energy

A pyramid of energy or pyramid of productivity shows the production or turnover (the rate at which energy or mass is transferred from one trophic level to the next) of biomass at each trophic level. Instead of showing a single snapshot in time, productivity pyramids show the flow of energy through the food chain. Typical units are grams per square meter per year or calories per square meter per year. As with the others, this graph shows producers at the bottom and higher trophic levels on top.

When an ecosystem is healthy, this graph produces a standard ecological pyramid. This is because in order for the ecosystem to sustain itself, there must be more energy at lower trophic levels than there is at higher trophic levels. This allows organisms on the lower levels to not only maintain a stable population, but also to transfer energy up the pyramid. The exception to this generalization is when portions of a food web are supported by inputs of resources from outside the local community. In small, forested streams, for example, the volume of higher levels is greater than could be supported by the local primary production.

When energy is transferred to the next trophic level, typically only 10% or 12% of it is used to build new biomass, becoming stored energy (the rest going to metabolic processes).

Advantages of the pyramid of energy as a representation:

- It takes account of the rate of production over a period of time.

- Two species of comparable biomass may have very different life spans. Thus a direct comparison of their total biomasses is misleading, but their productivity is directly comparable.

- The relative energy chain within an ecosystem can be compared using pyramids of energy; also different ecosystems can be compared.

- There are no inverted pyramids.

- The input of solar energy can be added.

Disadvantages of the pyramid of energy as a representation:

- The rate of biomass production of an organism is required, which involves measuring growth and reproduction through time.

- There is still the difficulty of assigning the organisms to a specific trophic level. As well as the organisms in the food chains there is the problem of assigning the decomposers and detritivores to a particular trophic level.

Pyramid of Biomass

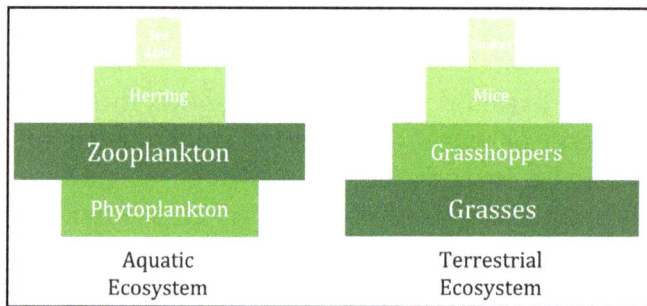

A pyramid of biomass shows the relationship between biomass and trophic level by quantifying the biomass present at each trophic level of an ecological community at a particular time. It is a graphical representation of biomass (total amount of living or organic matter in an ecosystem) present in unit area in different trophic levels. Typical units are grams per square meter, or calories per square meter. The pyramid of biomass may be "inverted". For example, in a pond ecosystem, the standing crop of phytoplankton, the major producers, at any given point will be lower than the mass of the heterotrophs, such as fish and insects. This is explained as the phytoplankton reproduce very quickly, but have much shorter individual lives.

Pyramid of Numbers

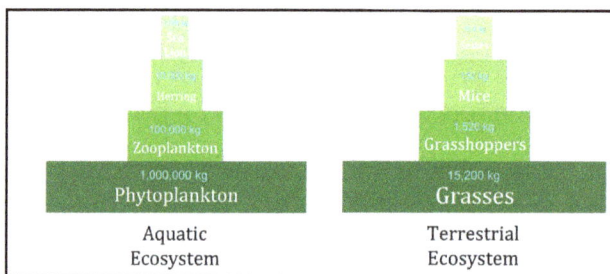

A pyramid of numbers shows the number of individual organisms involved at each trophic level in an ecosystem. The pyramids are not necessarily upright. In some ecosystems there can be more primary consumers than producers.

A pyramid of numbers shows graphically the population, or abundance, in terms of the number of individual organisms involved at each level in a food chain. This shows the number of organisms in each trophic level without any consideration for their individual sizes or biomass. The pyramid is not necessarily upright. For example, it will be inverted if beetles are feeding from the output of forest trees, or parasites are feeding on large host animals.

Ecological Buffers

Ecological buffers or Buffer zones are protected zones established around sensitive or critical areas — such as wildlife breeding or hibernation habitats, streams, and wetlands — to lessen the impacts of human activity and land disturbance. Well pads, roads and pipelines developed for shale oil and gas development reduce and fragment native forests, rivers and natural grasslands, reducing the quality of migration, foraging and nesting habitats for fish and wildlife. Changes in land cover can also have a negative impact on water quality and watershed health. Ecological buffers may be used to reduce or minimize the risks of land use disturbance and proximity of infrastructure specifically associated with shale energy development.

A variety of spatial patterns and arrangements for buffer zones exist, all following the same principle, but applied under completely different conditions (ecological, political, economic, etc). Hence, a wide diversity can be observed in the criteria for their creation and management. There are five aspects that are commonly considered in their creation. These are:

- Size: Determined based on factors such as the objectives for creation of buffer zone, availability of land, traditional land use systems, threats and opportunities.

- Ecology: Buffer zones vary depending on their focus on the landscape, habitat and species conservation, each of which demands a different approach for their creation.

- Economy: This involves appraisals such as cost-benefit analysis, time frame and discount rate, to assess economic viability of establishing a buffer zone.

- Legislation: Several international treaties and conventions (e.g. Convention on Biological Diversity, World Heritage Convention) and national level guidelines for protected areas (e.g. Nepal) recommend creation of buffer zones.

- Social and institutional: Creation of buffer zones also involves consideration of issues such as traditional rights of local communities, type of development activities to minimize negative impacts of conservation, local organisations to manage buffer zones and land tenure.

Management

There are various approaches in buffer zone management depending on the type and objectives of the conservation area for which they are created. For instance, activities in the buffer zones around some protected areas or World Heritage sites are recommended to be regulated so as to protect the core zone. In UNESCO Man and the Biosphere Reserves, socio-economic development of local communities play a crucial role. A buffer zone can also be managed as an area for research to develop approaches for sustainable use of resources, for ecosystem restoration, education and training, as well as carefully designed tourism and recreation activities. The degree of legal protection to buffer zone varies. In most cases where the buffer zones are outside the protected area, they fall under the institutional control and jurisdiction of authorities other than those responsible for management of the protected area.

References

- Systems-ecology, science: britannica.com, Retrieved 14 July, 2019

- Woodside, arch g.; caldwell, marylouise; spurr, ray (2006). "advancing ecological systems theory in lifestyle, leisure, and travel research". Journal of travel research. 44 (3): 259–272. Doi:10.1177/0047287505282945

- The-laws-of-thermodynamics-ecosystem-ecology, thermodynamics, ecosystem: ecologyessays. com, Retrieved 11 January, 2019

- Montes, f.; cañellas, i. (2006). "modelling coarse woody debris dynamics in even-aged scots pine forests". Forest ecology and management. 221 (1–3): 220–232. Doi:10.1016/j.foreco.2005.10.019

- Environmental-studies-energy-flow-in-ecosystem, environmental-studies: tutorialspoint.com Retrieved 18 April, 2019

- Stockdale, e. A.; shepherd, m. A.; fortune, s.; cuttle, s. P. (2006). "soil fertility in organic farming systems – fundamentally different?". Soil use and management. 18 (s1): 301–308. Doi:10.1111/j.1475-2743.2002.tb00272.x

PERMISSIONS

INDEX

A

Abiotic Components, 39-41, 61, 167

Acid Rain, 44, 86, 121-123, 135, 152, 154, 166, 182

Alpine Tundra, 73, 88, 91-92

Amino Acids, 13

Aquatic Ecosystem, 36, 65-66, 151-152

Auxin, 99-100

B

Behavioral Ecology, 1, 3, 5, 30

Biogeochemical Cycling, 8, 11-12

Biological Diversity, 8, 13, 15, 57, 59, 79, 181, 217

Biomass, 3, 6, 15, 43, 54, 87, 93, 121, 123, 133, 162, 166, 193, 195, 200, 207, 209-210, 212, 214-217

Biomes, 43, 50, 61, 79, 187

Biotic Community, 15, 41, 55, 180

Biotic Components, 38-39, 64, 172

Boreal Forest, 46-47, 72-75, 77-78, 80-87, 94

C

Cactoblastis Cactorum, 26

Carbon Cycle, 12, 121, 135, 197-198

Cellulose, 43, 201

Checkerspot Butterfly, 29

Climate Change, 8, 49, 60, 68, 84, 96, 116, 135

Climax Community, 33-34

Community Ecology, 5-6

Coniferous Trees, 44-45, 80

Coral Reef, 35

Cyanobacteria, 9-10, 12, 206

D

Deciduous Forests, 34, 42, 45-47, 77

Decomposition, 8, 10, 54-55, 68, 87, 162, 171, 176, 180-181, 196-197, 199, 202-203, 206

Deforestation, 48, 55, 59-60, 119

Desert Ecosystem, 36, 60

E

Ecological Succession, 1, 33-35, 55

Ecosystems, 2-3, 6-11, 14-15, 31-33, 35-36, 39-40, 46, 49, 53, 61, 69, 72, 93, 115, 118, 121, 123-124, 167, 170, 174, 180, 183-189, 193, 206-207, 216

Energy Budget, 37

Euphydryas Editha, 29

Evaporation Rate, 64

F

Food Chain, 6, 10, 37-38, 152, 162, 189, 195-196, 200, 204-206, 208-209, 215, 217

Food Web, 10-11, 37, 80, 183, 186, 189, 200-201, 204-207, 209-215

Forest Ecology, 1, 14-15, 44, 53, 218

Forest Tundra, 46

G

Genetic Variation, 16-17, 58

Geospiza Scandens, 21

Greenhouse Gas, 6, 122, 126

H

Habitat Fragmentation, 49, 124

Hedgehogs, 30, 65

Heterotrophs, 10, 37-38, 205-207, 216

Hydrosphere, 14, 31, 134, 196

I

Intertidal Zone, 65-66

L

Lentic Ecosystems, 66, 154, 156-157, 167

Lichen Woodland, 46, 78, 87-88

Lotic Ecosystems, 65, 154, 156, 167, 180

M

Methane Metabolism, 8, 12

Montane Ecosystem, 93

N

Natural Selection, 16-17

Nitrogen Fixation, 8, 11, 13, 133

Nutrient Cycling, 3, 6-7, 15, 132-133, 136, 200, 202, 210

P
Photorespiration, 40, 108-109
Photosynthesis, 3, 6, 8, 10, 12, 36, 38, 64, 79, 95, 98, 100, 104-110, 153, 156, 158-159, 162, 169, 171, 176, 181, 185, 194-197, 205, 207
Phytoplankton, 6, 152-153, 159-160, 163-166, 173, 180, 183, 185-186, 210, 216
Plant Ecology, 95
Plant Perception, 95, 97, 102
Plant Stress, 95, 103-109
Polar Climate, 91
Population Cycles, 25
Population Ecology, 1, 5, 15, 29, 137, 203
Potential Evapotranspiration, 53, 64, 93
Precipitation, 11, 47-49, 54-55, 63, 68, 75-76, 79, 82, 85-86, 89, 93, 122, 125, 168, 170, 179, 182

S
Savannas, 46, 67, 70-71
Scots Pine, 80, 218

Shrublands Biome, 67, 70-72
Soil Fertility, 58, 68, 218
Soil Organic Matter, 58
Stratification, 53, 157-158
Subalpine Forest, 73
Sulfur Metabolism, 8, 12
Sustainable Forest Management, 44

T
Taiga, 36, 42-43, 46, 72-90
Temperate Forest, 47, 76, 85
Terrestrial Ecology, 1, 7
Theoretical Ecology, 31, 33
Topography, 15, 77, 96, 200
Tropical Dry Forests, 46
Tropical Rainforests, 42, 48-49, 51, 53-56, 60

W
Water Budget, 46
Wetlands, 65-66, 87, 90, 151, 154, 156, 165-167, 205, 207, 217
White Spruce, 73, 82-83, 85-86